Almuric

Robert E. Howard

IAP © 2009

Printed in Scotts Valley, CA – USA.

Howard, Robert E.

Almuric / Robert E. Howard– 1st ed.

   1.   Literature – fiction

Cover Images

© Aliencat | Dreamstime.com

© Prometeus | Dreamstime.com

FOREWORD

It was not my original intention ever to divulge the whereabouts of Esau Cairn, or the mystery surrounding him. My change of mind was brought about by Cairn himself, who retained a perhaps natural and human desire to give his strange story to the world which had disowned him and whose members can now never reach him. What he wishes to tell is his affair. One phase of my part of the transaction I refuse to divulge; I will not make public the means by which I transported Esau Cairn from his native Earth to a planet in a solar system undreamed of by even the wildest astronomical theorists. Nor will I divulge by what means I later achieved communication with him, and heard his story from his own lips, whispering ghostly across the cosmos.

Let me say that it was not premeditated. I stumbled upon the Great Secret quite by accident in the midst of a scientific experiment, and never thought of putting it to any practical use, until that night when Esau Cairn groped his way into my darkened observatory, a hunted man, with the blood of a human being on his hands. It was chance led him there, the blind instinct of the hunted thing to find a den wherein to turn at bay.

Let me state definitely and flatly, that, whatever the appearances against him, Esau Cairn is not, and was never, a criminal. In that specific case he was merely the pawn of a corrupt political machine which turned on him when he realized his position and refused to comply further with its demands. In general, the acts of his life which might suggest a violent and unruly nature simply sprang from his peculiar mental make-up.

Science is at last beginning to perceive that there is sound truth in the popular phrase, "born out of his time." Certain natures are attuned to certain phases or epochs of history, and these natures, when cast by chance into an age alien to their reactions and emotions, find difficulty

3

in adapting themselves to their surroundings. It is but another example of nature's inscrutable laws, which sometimes are thrown out of stride by some cosmic friction or rift, and result in havoc to the individual and the mass.

Many men are born outside their century; Esau Cairn was born outside his epoch. Neither a moron nor a low-class primitive, possessing a mind well above the average, he was, nevertheless, distinctly out of place in the modern age. I never knew a man of intelligence so little fitted for adjustment in a machine-made civilization. (Let it be noted that I speak of him in the past tense; Esau Cairn lives, as far as the cosmos is concerned; as far as the Earth is concerned, he is dead, for he will never again set foot upon it.)

He was of a restless mold, impatient of restraint and resentful of authority. Not by any means a bully, he at the same time refused to countenance what he considered to be the slightest infringement on his rights. He was primitive in his passions, with a gusty temper and a courage inferior to none on this planet. His life was a series of repressions. Even in athletic contests he was forced to hold himself in, lest he injure his opponents. Esau Cairn was, in short, a freak-a man whose physical body and mental bent leaned back to the primordial.

Born in the Southwest, of old frontier stock, he came of a race whose characteristics were inclined toward violence, and whose traditions were of war and feud and battle against man and nature. The mountain country in which he spent his boyhood carried out the tradition. Contest-physical contest-was the breath of life to him. Without it he was unstable and uncertain. Because of his peculiar physical make-up, full enjoyment in a legitimate way, in the ring or on the football field was denied him. His career as a football player was marked by crippling injuries received by men playing against him, and he was branded as an unnecessarily brutal man, who fought to maim his opponents rather than win games. This was unfair. The injuries were simply resulting from the use of his great strength, always so far superior to that of the men opposed to him. Cairn was not a great sluggish lethargic giant as so many powerful men are; he was vibrant with fierce life, ablaze with dynamic energy. Carried away by the lust of combat, he forgot to control his powers, and the result was broken limbs or fractured skulls for his opponents.

It was for this reason that he withdrew from college life, unsatisfied and embittered, and entered the professional ring. Again his fate dogged him. In his training-quarters, before he had had a single match, he almost fatally injured a sparring partner. Instantly the papers pounced upon the incident, and played it up beyond its natural proportions. As a result Cairn's license was revoked.

Bewildered, unsatisfied, he wandered over the world, a restless Hercules, seeking outlet for the immense vitality that surged tumultuously within him, searching vainly for some form of life wild and strenuous enough to satisfy his cravings, born in the dim red days of the world's youth.

Of the final burst of blind passion that banished him for ever from the life wherein he roamed, a stranger, I need say little. It was a nine-days' wonder, and the papers exploited it with screaming headlines. It was an old story-a rotten city government, a crooked political boss, a man chosen, unwittingly on his part, to be used as a tool and serve as a puppet.

Cairn, restless, weary of the monotony of a life for which he was unsuited, was an ideal tool-for a while. But Cairn was neither a criminal nor a fool. He understood their game quicker than they expected, and took a stand surprisingly firm to them, who did not know the real man.

Yet, even so, the result would not have been so violent if the man who had used and ruined Cairn had any real intelligence. Used to grinding men under his feet and seeing them cringe and beg for mercy, Boss Blaine could not understand that he was dealing with a man to whom his power and wealth meant nothing. Yet so schooled was Cairn to iron self-control that it required first a gross insult, then an actual blow on

5

the part of Blaine, to rouse him. Then for the first time in his life, his wild nature blazed into full being. All his thwarted and repressed life surged up behind the clenched fist that broke Blaine's skull like an eggshell and stretched him lifeless on the floor, behind the desk from which he had for years ruled a whole district.

Cairn was no fool. With the red haze of fury fading from his glare, he realized that he could not hope to escape the vengeance of the machine that controlled the city. It was not because of fear that he fled Blaine's house. It was simply because of his primitive instinct to find a more convenient place to turn at bay and fight out his death fight.

So it was that chance led him to my observatory.

He would have left, instantly, not wishing to embroil me in his trouble, but I persuaded him to remain and tell me his story. I had long expected some catastrophe of the sort. That he had repressed himself as long as he did, shows something of his iron character. His nature was as wild and untamed as that of a manned lion.

He had no plan-he simply intended to fortify himself somewhere and fight it out with the police until he was riddled with lead.

I at first agreed with him, seeing no better alternative. I was not so naive as to believe he had any chance in the courts with the evidence that would be presented against him. Then a sudden thought occurred to me, so fantastic and alien, and yet so logical, that I instantly propounded it to my companion. I told him of the Great Secret, and gave him proof of its possibilities.

In short, I urged him to take the chance of a flight through space, rather than meet the certain death that awaited him.

And he agreed. There was no place in the universe which would support human life. But I had looked beyond the knowledge of men, in universes beyond universes. And I chose the only planet I knew on which a human being could exist-the wild, primitive, and strange planet I named Almuric.

Cairn understood the risks and uncertainties as well as I. But he was utterly fearless-and the thing was done. Esau Cairn left the planet of his birth, for a world swimming afar in space, alien, aloof, strange.

Esau Cairn's Narrative

CHAPTER 1

The Transition was so swift and brief, that it seemed less than a tick of time lay between the moment I placed myself in Professor Hildebrand's strange machine, and the instant when I found myself standing upright in the clear sunlight that flooded a broad plain. I could not doubt that I had indeed been transported to another world. The landscape was not so grotesque and fantastic as I might have supposed, but it was indisputably alien to anything existing on the Earth.

But before I gave much heed to my surroundings, I examined my own person to learn if I had survived that awful flight without injury. Apparently I had. My various parts functioned with their accustomed vigor. But I was naked. Hildebrand had told me that inorganic substance could not survive the transmutation. Only vibrant, living matter could pass unchanged through the unthinkable gulfs which lie between the planets. I was grateful that I had not fallen into a land of ice and snow. The plain seemed filled with a lazy summer-like heat. The warmth of the sun was pleasant on my bare limbs.

On every side stretched away a vast level plain, thickly grown with short green grass. In the distance this grass attained a greater height, and through it I caught the glint of water. Here and there throughout the plain this phenomenon was repeated, and I traced the meandering course of several rivers, apparently of no great width. Black dots moved through the grass near the rivers, but their nature I could not determine. However, it was quite evident that my lot had not been cast on an uninhabited planet, though I could not guess the nature of the inhabitants. My imagination peopled the distances with nightmare shapes.

It is an awesome sensation to be suddenly hurled from one's native world into a new strange alien sphere. To say that I was not appalled at the prospect, that I did not shrink and shudder in spite of the peaceful quiet of my environs, would be hypocrisy. I, who had never known fear, was transformed into a mass of quivering, cowering nerves, starting at my own shadow. It was that man's utter helplessness was borne in upon me, and my mighty frame and massive thews seemed frail and brittle as the body of a child. How could I pit them against an unknown world? In that instant I would gladly have returned to Earth and the gallows that awaited me, rather than face the nameless terrors with which imagination peopled my new-found world. But I was soon to learn that those thews I now despised were capable of carrying me through greater perils than I dreamed.

A slight sound behind me brought me around to stare amazedly at the first inhabitant of Almuric I was to encounter. And the sight, awesome and menacing as it was, yet drove the ice from my veins and brought back some of my dwindling courage. The tangible and material can never be as grisly as the unknown, however perilous.

At my first startled glance I thought it was a gorilla which stood before me. Even with the thought I realized that it was a man, but such a man as neither I nor any other Earthman had ever looked upon.

He was not much taller than I, but broader and heavier, with a great spread of shoulders, and thick limbs knotted with muscles. He wore a loincloth of some silk-like material girdled with a broad belt which supported a long knife in a leather sheath. High-strapped sandals were on his feet. These details I took in at a glance, my attention being instantly fixed in fascination on his face.

Such a countenance it is difficult to imagine or describe. The head was set squarely between the massive shoulders, the neck so squat as to be scarcely apparent. The jaw was square and powerful, and as the wide thin lips lifted in a snarl, I glimpsed brutal tusk-like teeth. A short bristly beard masked the jaw, set off by fierce, up-curving mustaches. The nose was almost rudimentary, with wide flaring nostrils. The eyes were small, bloodshot, and an icy gray in color. From the thick black brows the forehead, low and receding, sloped back into a tangle of coarse, bushy hair. The ears were small and very close-set.

The mane and beard were very blue-black, and the creature's limbs and body were almost covered with hair of the same hue. He was not, indeed, as hairy as an ape, but he was hairier than any human being I had ever seen.

I instantly realized that the being, hostile or not, was a formidable figure. He fairly emanated strength-hard, raw, brutal power. There was not an ounce of surplus flesh on him. His frame was massive, with heavy bones. His hairy skin rippled with muscles that looked iron-hard. Yet it was not altogether his body that spoke of dangerous power. His look, his carriage, his whole manner reflected a terrible physical might backed by a cruel and implacable mind. As I met the blaze of his bloodshot eyes, I felt a wave of corresponding anger. The stranger's attitude was arrogant and provocative beyond description. I felt my muscles tense and harden instinctively.

But for an instant my resentment was submerged by the amazement with which I heard him speak in perfect English!

"Thak! What manner of man are you?"

His voice was harsh, grating and insulting. There was nothing subdued or restrained about him. Here were the naked primitive instincts and manners, unmodified. Again I felt the old red fury rising in me, but I fought it down.

"I am Esau Cairn," I answered shortly, and halted, at a loss how to explain my presence on his planet.

His arrogant eyes roved contemptuously over my hairless limbs and smooth face, and when he spoke, it was with unbearable scorn.

"By Thak, are you a man or a woman?"

My answer was a smash of my clenched fist that sent him rolling on the sward.

The act was instinctive. Again my primitive wrath had betrayed me. But I had no time for self-reproach. With a scream of bestial rage my enemy sprang up and rushed at me, roaring and frothing. I met him breast to breast, as reckless in my wrath as he, and in an instant was fighting for my life.

I, who had always had to restrain and hold down my strength lest I injure my fellow men, for the first time in my life found myself in the clutches of a man stronger than myself. This I realized in the first instant of impact, and it was only by the most desperate efforts that I fought clear of his crushing embrace.

The fight was short and deadly. The only thing that saved me was the fact that my antagonist knew nothing of boxing. He could-and did- strike powerful blows with his clenched fists, but they were clumsy, ill-timed and erratic. Thrice I mauled my way out of grapples that would have ended with the snapping of my spine. He had no knack of avoiding blows; no man on Earth could have survived the terrible battering I gave him. Yet he incessantly surged in on me, his mighty hands spread to drag me down. His nails were almost like talons, and I was quickly bleeding from a score of places where they had torn the skin.

Why he did not draw his dagger I could not understand, unless it was because he considered himself capable of crushing me with his bare hands- which proved to be the case. At last, half blinded by my smashes, blood gushing from his split ears and splintered teeth, he did reach for his weapon, and the move won the fight for me.

Breaking out of a half-clinch, he straightened out of his defensive crouch and drew his dagger. And as he did so, I hooked my left into his belly with all the might of my heavy shoulders and powerfully driving legs behind it. The breath went out of him in an explosive gasp, and my fist sank to the wrist in his belly. He swayed, his mouth flying open, and I smashed my right to his sagging jaw. The punch started at my hip, and carried every ounce of my weight and strength. He went down like a slaughtered ox and lay without twitching, blood spreading out over his beard. That last smash had torn his lip open from the corner of his mouth to the rim of his chin, and must surely have fractured his jawbone as well.

Panting from the fury of the bout, my muscles aching from his crushing grasp, I worked my raw, skinned knuckles, and stared down at my victim, wondering if I had sealed my doom. Surely, I could expect nothing now but hostility from the people of Almuric. Well, I thought, as well be hanged for a sheep as a goat. Stooping, I despoiled my adversary of his single garment, belt and weapon, and transferred them to my own frame. This done, I felt some slight renewal of confidence. At least I was partly clothed and armed.

I examined the dagger with much interest. A more murderous weapon I have never seen. The blade was perhaps nineteen inches in length, double-edged, and sharp as a razor. It was broad at the haft, tapering to a diamond point. The guard and pommel were of silver, the hilt covered with a substance somewhat like shagreen. The blade was indisputably steel, but of a quality I had never before encountered. The whole was a triumph of the weapon-maker's art, and seemed to indicate a high order of culture.

From my admiration of my newly acquired weapon, I turned again to my victim, who was beginning to show signs of returning consciousness. Instinct caused me to sweep the grasslands, and in the distance, to the south, I saw a group of figures moving toward me. They were surely men, and armed men. I caught the flash of the sunlight on steel. Perhaps they were of the tribe of my adversary. If they found me standing over their senseless comrade, wearing the spoils of conquest, their attitude toward me was not hard to visualize.

I cast my eyes about for some avenue of escape or refuge, and saw that the plain, some distance away, ran up into low green-clad foothills. Beyond these in turn, I saw larger hills, marching up and up in serried ranges. Another glance showed the distant figures to have vanished among the tall grass along one of the river courses, which they must cross before they reached the spot where I stood.

Waiting for no more, I turned and ran swiftly toward the hills. I did not lessen my pace until I reached the foot of the first foothills, where I ventured to look back, my breath coming in gasps, and my heart pounding suffocatingly from my exertions. I could see my antagonist, a small shape in the vastness of the plain. Further on, the group I was seeking to avoid had come into the open and were hastening toward him.

I hurried up the low slope, drenched with sweat and trembling with fatigue. At the crest I looked back once more, to see the figures clustered about my vanquished opponent. Then I went down the opposite slope quickly, and saw them no more.

An hour's journeying brought me into as rugged a country as I have ever seen. On all sides rose steep slopes, littered with loose boulders, which threatened to roll down upon the wayfarer. Bare stone cliffs, reddish in color, were much in evidence. There was little vegetation, except for low stunted trees, of which the spread of their branches was equal to the height of the trunk, and several varieties of thorny bushes, upon some of which grew nuts of peculiar shape and color. I broke open several of these, finding the kernel to be rich and meaty in appearance, but I dared not eat it, although I was feeling the bite of hunger.

My thirst bothered me more than my hunger, and this at least I was able to satisfy, although the satisfying nearly cost me my life. I clambered down a precipitous steep and entered a narrow valley, enclosed by lofty cliffs, at the foot of which the nut-bearing bushes grew in great abundance. In the middle of the valley lay a broad pool, apparently fed by a spring. In the center of the pool the water bubbled continuously, and a small stream led off down the valley.

I approached the pool eagerly, and lying on my belly at its lush-grown marge, plunged my muzzle into the crystal-clear water. It, too, might be lethal for an Earthman, for all I knew, but I was so maddened with thirst that I risked it. It had an unusual tang, a quality I have always found present in Almuric water, but it was deliciously cold and satisfying. So pleasant it was to my parched lips that after I had satisfied my thirst, I lay there enjoying the sensation of tranquility. That was a mistake. Eat quickly, drink quickly, sleep lightly, and linger not over anything-those are the first rules of the wild, and his life is not long who fails to observe them.

The warmth of the sun, the bubbling of the water, the sensuous feeling of relaxation and satiation after fatigue and thirst-these wrought on me like an opiate to lull me into semis-lumber. It must have been some subconscious instinct that warned me, when a faint swishing reached my ears that was not part of the rippling of the spring. Even before my mind translated the sound as the passing of a heavy body through the tall grass, I whirled on my side, snatching at my poniard.

Simultaneously my ears were stunned with a deafening roar, there was a rushing through the air, and a giant form crashed down where I had lain

12

an instant before, so close to me that its outspread talons raked my thigh. I had no time to tell the nature of my attacker-I had only a dazed impression that it was huge, supple, and catlike. I rolled frantically aside as it spat and struck at me sidewise; then it was on me, and even as I felt its claws tear agonizingly into my flesh, the ice-cold water engulfed us both. A catlike yowl rose half strangled, as if the yowler had swallowed a large amount of water. There was a great splashing and thrashing about me; then as I rose to the surface, I saw a long, bedraggled shape disappearing around the bushes near the cliffs. What it was I could not say, but it looked more like a leopard than anything else, though it was bigger than any leopard I had ever seen.

Scanning the shore carefully, I saw no other enemy, and crawled out of the pool, shivering from my icy plunge. My poniard was still in its scabbard. I had had no time to draw it, which was just as well. If I had not rolled into the pool, just when I did, dragging my attacker with me, it would have been my finish. Evidently the beast had a true catlike distaste for water.

I found that I had a deep gash in my thigh and four lesser abrasions on my shoulder, where a great talon-armed paw had closed. The gash in my leg was pouring blood, and I thrust the limb deep into the icy pool, swearing at the excruciating sting of the cold water on the raw flesh. My leg was nearly numb when the bleeding ceased.

I now found myself in a quandary. I was hungry, night was coming on, there was no telling when the leopard-beast might return, or another predatory animal attack me; more than that, I was wounded. Civilized man is soft and easily disabled. I had a wound such as would be considered, among civilized people, ample reason for weeks of an invalid's existence. Strong and rugged as I was, according to Earth standards, I despaired when I surveyed the wound, and wondered how I was to treat it. The matter was quickly taken out of my hands.

I had started across the valley toward the cliffs, hoping I might find a cave there, for the nip of the air warned me that the night would not be as warm as the day, when a hellish clamor up near the mouth of the valley caused me to wheel and glare in that direction. Over the ridge came what I thought to be a pack of hyenas, except for their noise, which was more infernal than an Earthly hyena, even, could produce. I had no illusions as to their purpose. It was I they were after.

Necessity recognizes few limitations. An instant before I had been limping painfully and slowly. Now I set out on a mad race for the cliff as if I were fresh and unwounded. With every step a spasm of agony shot along my thigh, and the wound, bleeding afresh, spurted red, but I gritted my teeth and increased my efforts.

13

My pursuers gave tongue and raced after me with such appalling speed that I had almost given up hope of reaching the trees beneath the cliffs before they pulled me down. They were snapping at my heels when I lurched into the low stunted growths, and swarmed up the spreading branches with a gasp of relief. But to my horror the hyenas climbed after me! A desperate downward glance showed me that they were not true hyenas; they differed from the breed I had known just as everything on Almuric differed subtly from its nearest counterpart on Earth. These beasts had curving catlike claws, and their bodily structure was catlike enough to allow them to climb as well as a lynx.

Despairingly, I was about to turn at bay, when I saw a ledge on the cliff above my head. There the cliff was deeply weathered, and the branches pressed against it. A desperate scramble up the perilous slant, and I had dragged my scratched and bruised body up on the ledge and lay glaring down at my pursuers, who loaded the topmost branches and howled up at me like lost souls. Evidently their climbing ability did not include cliffs, because after one attempt, in which one sprang up toward the ledge, clawed frantically for an instant on the sloping stone wall, and then fell off with an awful shriek, they made no effort to reach me.

Neither did they abandon their post. Stars came out, strange unfamiliar constellations that blazed whitely in the dark velvet skies, and a broad golden moon rose above the cliffs, and flooded the hills with weird light; but still my sentinels sat on the branches below me and howled up at me their hatred and belly-hunger.

The air was icy, and frost formed on the bare stone where I lay. My limbs became stiff and numb. I had knotted my girdle about my leg for a tourniquet; the run had apparently ruptured some small veins laid bare by the wound, because the blood flowed from it in an alarming manner.

I never spent a more miserable night. I lay on the frosty stone ledge, shaking with cold. Below me the eyes of my hunters burned up at me. Throughout the shadowy hills sounded the roaring and bellowing of unknown monsters. Howls, screams and yapping cut the night. And there I lay, naked, wounded, freezing, hungry, terrified, just as one of my remote ancestors might have lain in the Paleolithic Age of my own planet.

I can understand why our heathen ancestors worshipped the sun. When at last the cold moon sank and the sun of Almuric pushed its golden rim above the distant cliffs, I could have wept for sheer joy. Below me the hyenas snarled and stretched themselves, bayed up at me briefly, and loped away in search of easier prey. Slowly the warmth of the sun stole through my cramped, numbed limbs, and I rose stiffly up to greet the day, just as that forgotten forbear of mine might have stood up in the youth dawn of

14

the Earth.

After a while I descended, and fell upon the nuts clustered in the bushes near by. I was faint from hunger, and decided that I had as soon die from poisoning as from starvation. I broke open the thick shells and munched the meaty kernels eagerly, and I cannot recall any Earthly meal, howsoever elaborate, that tasted half as good. No ill effects followed; the nuts were good and nutritious. I was beginning to overcome my surroundings, at least so far as food was concerned. I had surmounted one obstacle of life on Almuric.

It is needless for me to narrate the details of the following months. I dwelt among the hills in such suffering and peril as no man on Earth has experienced for thousands of years. I make bold to say that only a man of extraordinary strength and ruggedness could have survived as I did. I did more than survive. I came at last to thrive on the existence.

At first I dared not leave the valley, where I was sure of food and water. I built a sort of nest of branches and leaves on the ledge, and slept there at night. Slept? The word is misleading. I crouched there, trying to keep from freezing, grimly lasting out the night. In the daytime I snatched naps, learning to sleep anywhere, or at any time, and so lightly that the slightest unusual noise would awaken me. The rest of the time I explored my valley and the hills about, and picked and ate nuts. Nor were my humble explorations uneventful. Time and again I raced for the cliffs or the trees, winning sometimes by shuddering hairbreadths. The hills swarmed with beasts, and all seemed predatory.

It was that fact which held me to my valley, where I at least had a bit of safety. What drove me forth at last was the same reason that has always driven forth the human race, from the first ape-man down to the last European colonist-the search for food. My supply of nuts became exhausted. The trees were stripped. This was not altogether on my account, although I developed a most ravenous hunger, what of my constant exertions; but others came to eat the nuts-huge shaggy bear-like creatures, and things that looked like fur-clad baboons. These animals ate nuts, but they were omnivorous, to judge by the attention they accorded me. The bears were comparatively easy to avoid; they were mountains of flesh and muscle, but they could not climb, and their eyes were none too good. It was the baboons I learned to fear and hate. They pursued me on sight, they could both run and climb, and they were not balked by the cliff.

One pursued me to my eyrie, and swarmed up onto the ledge with me. At least such was his intention, but man is always most dangerous when cornered. I was weary of being hunted. As the frothing apish monstrosity hauled himself up over my ledge, manlike, I drove my poniard down

15

between his shoulders with such fury that I literally pinned him to the ledge; the keen point sinking a full inch into the solid stone beneath him.

The incident showed me both the temper of my steel, and the growing quality of my own muscles. I who had been among the strongest on my own planet, found myself a weakling on primordial Almuric. Yet the potentiality of mastery was in my brain and my thews, and I was beginning to find myself.

Since survival was dependent on toughening, I toughened. My skin, burnt brown by the sun and hardened by the elements, became more impervious to both heat and cold than I had deemed possible. Muscles I had not known I possessed became evident. Such strength and suppleness became mine as Earthmen have not known for ages.

A short time before I had been transported from my native planet, a noted physical culture expert had pronounced me the most perfectly developed man on Earth. As I hardened with my fierce life on Almuric, I realized that the expert honestly had not known what physical development was. Nor had I. Had it been possible to divide my being and set opposite each other the man that expert praised, and the man I had become, the former would have seemed ridiculously soft, sluggish and clumsy in comparison to the brown, sinewy giant opposed to him.

I no longer turned blue with the cold at night, nor did the rockiest way bruise my naked feet. I could swarm up an almost sheer cliff with the ease of a monkey, I could run for hours without exhaustion; in short dashes it would have taken a racehorse to outfoot me. My wounds, untended except for washing in cold water, healed of themselves, as Nature is prone to heal the hurts of such as live close to her.

All this I narrate in order that it may be seen what sort of a man was formed in the savage mold. Had it not been for the fierce forging that made me steel and rawhide, I could not have survived the grim bloody episodes through which I was to pass on that wild planet.

With new realization of power came confidence. I stood on my feet and stared at my bestial neighbors with defiance. I no longer fled from a frothing, champing baboon. With them, at least, I declared feud, growing to hate the abominable beasts as I might have hated human enemies. Besides, they ate the nuts I wished for myself.

They soon learned not to follow me to my eyrie, and the day came when I dared to meet one on even terms, I will never forget the sight of him frothing and roaring as he charged out of a clump of bushes, and the awful glare in his manlike eyes. My resolution wavered, but it was too late to retreat, and I met him squarely, skewering him through the heart as he closed in with his long clutching arms.

But there were other beasts which frequented the valley, and which I did not attempt to meet on any terms: the hyenas, the saber-tooth leopards, longer and heavier than an Earthly tiger and more ferocious; giant moose-like creatures, carnivorous, with alligator-like tusks; the monstrous bears; gigantic boars, with bristly hair which looked impervious to a swordcut. There were other monsters, which appeared only at night, and the details of which I was not able to make out. These mysterious beasts moved mostly in silence, though some emitted high-pitched weird wails, or low Earth-shaking rumbles. As the unknown is most menacing, I had a feeling that these nighted monsters were even more terrible than the familiar horrors which harried my day-life.

I remember one occasion on which I awoke suddenly and found myself lying tensely on my ledge, my ears strained to a night suddenly and breathlessly silent. The moon had set and the valley was veiled in darkness. Not a chattering baboon, not a yelping hyena disturbed the sinister stillness. *Something* was moving through the valley; I heard the faint rhythmic swishing of the grass that marked the passing of some huge body, but in the darkness I made out only a dim gigantic shape, which somehow seemed infinitely longer than it was broad-out of natural proportion, somehow. It passed away up the valley, and with its going, it was as if the night audibly expelled a gusty sigh of relief. The nocturnal noises started up again, and I lay back to sleep once more with a vague feeling that some grisly horror had passed me in the night.

I have said that I strove with the baboons over the possession of the life-giving nuts. What of my own appetite and those of the beasts, there came a time when I was forced to leave my valley and seek far afield in search of nutriment. My explorations had become broader and broader, until I had exhausted the resources of the country close about. So I set forth at random through the hills in a southerly and easterly direction. Of my wanderings I will deal briefly. For many weeks I roamed through the hills, starving, feasting, threatened by savage beasts sleeping in trees or perilously on tall rocks when night fell. I fled, I fought, I slew, I suffered wounds. Oh, I can tell you my life was neither dull nor uneventful.

I was living the life of the most primitive savage; I had neither companionship, books, clothing, nor any of the things which go to make up civilization. According to the cultured viewpoint, I should have been most miserable. I was not. I reveled in my existence. My being grew and expanded. I tell you, the natural life of mankind is a grim battle for existence against the forces of nature, and any other form of life is artificial and without realistic meaning.

17

My life was not empty; it was crowded with adventures calling on every ounce of intelligence and physical power. When I swung down from my chosen eyrie at dawn, I knew that I would see the sun set only through my personal craft and strength and speed. I came to read the meaning of every waving grass tuft, each masking bush, each towering boulder. On every hand lurked Death in a thousand forms. My vigilance could not be relaxed, even in sleep. When I closed my eyes at night it was with no assurance that I would open them at dawn. I was fully alive. That phrase has more meaning than appears on the surface. The average civilized man is never fully alive; he is burdened with masses of atrophied tissue and useless matter. Life flickers feebly in him; his senses are dull and torpid. In developing his intellect he has sacrificed far more than he realizes.

I realized that I, too, had been partly dead on my native planet. But now I was alive in every sense of the word; I tingled and burned and stung with life to the finger tips and the ends of my toes. Every sinew, vein, and springy bone was vibrant with the dynamic flood of singing, pulsing, humming life. My time was too much occupied with food-getting and preserving my skin to allow the developing of the morbid and intricate complexes and inhibitions which torment the civilized individual. To those highly complex persons who would complain that the psychology of such a life is over-simple, I can but reply that in my life at that time, violent and continual action and the necessity of action crowded out most of the gropings and soul-searchings common to those whose safety and daily meals are assured them by the toil of others. My life *was* primitively simple; I dwelt altogether in the present. My life on Earth already seemed like a dream, dim and far away.

All my life I had held down my instincts, had chained and enthralled my over-abundant vitalities. Now I was free to hurl all my mental and physical powers into the untamed struggle for existence, and I knew such zest and freedom as I had never dreamed of.

In all my wanderings-and since leaving the valley I had covered an enormous distance-I had seen no sign of humanity, or anything remotely resembling humanity.

It was the day I glimpsed a vista of rolling grassland beyond the peaks, that I suddenly encountered a human being. The meeting was unexpected. As I strode along an upland plateau, thickly grown with bushes and littered with boulders, I came abruptly on a scene striking in its primordial significance.

Ahead of me the Earth sloped down to form a shallow bowl, the floor of which was thickly grown with tall grass, indicating the presence of a spring. In the midst of this bowl a figure similar to the one I had

18

encountered on my arrival on Almuric was waging an unequal battle with a saber-tooth leopard. I stared in amazement, for I had not supposed that any human could stand before the great cat and live.

Always the glittering wheel of a sword shimmered between the monster and its prey, and blood on the spotted hide showed that the blade had been fleshed more than once. But it could not last; at any instant I expected to see the swordsman go down beneath the giant body.

Even with the thought, I was running fleetly down the shallow slope. I owed nothing to the unknown man, but his valiant battle stirred newly plumbed depths in my soul. I did not shout but rushed in silently and murderously, my poniard gleaming in my hand. Even as I reached them, the great cat sprang, the sword went spinning from the wielder's hand, and he went down beneath the hurtling bulk. And almost simultaneously I disemboweled the saber-tooth with one tremendous ripping stroke.

With a scream it lurched off its victim, slashing murderously as I leaped back, and then it began rolling and tumbling over the grass, roaring hideously and ripping up the Earth with its frantic talons, in a ghastly welter of blood and streaming entrails.

It was a sight to sicken the hardiest, and I was glad when the mangled beast stiffened convulsively and lay still.

I turned to the man, but with little hope of finding life in him. I had seen the terrible saber-like fangs of the giant carnivore tear into his throat as he went down.

He was lying in a wide pool of blood, his throat horribly mangled. I could see the pulsing of the great jugular vein which had been laid bare, though not severed. One of the huge taloned paws had raked down his side from arm-pit to hip, and his thigh had been laid open in a frightful manner; I could see the naked bone, and from the ruptured veins blood was gushing. Yet to my amazement the man was not only living, but conscious. Yet even as I looked, his eyes glazed and the light faded in them.

I tore a strip from his loincloth and made a tourniquet about his thigh which somewhat slackened the flow of blood; then I looked down at him helplessly. He was apparently dying, though I knew something of the stamina and vitality of the wild and its people. And such evidently this man was; he was as savage and hairy in appearance, though not quite so bulky, as the man I had fought during my first day on Almuric.

As I stood there helplessly, something whistled venomously past my ear and thudded into the slope behind me. I saw a long arrow quivering there, and a fierce shout reached my ears. Glaring about, I saw half a dozen hairy men running fleetly toward me, fitting shafts to their bows as they came.

With an instinctive snarl I bounded up the short slope, the whistle of the missiles about my head lending wings to my heels. I did not stop, once I had gained the cover of the bushes surrounding the bowl, but went straight on, wrathful and disgusted. Evidently men as well as beasts were hostile on Almuric, and I would do well to avoid them in the future.

Then I found my anger submerged in a fantastic problem. I had understood some of the shouts of the men as they rushed toward me. The words had been in English, just as the antagonist of my first encounter had spoken and understood that language. In vain I cudgeled my mind for a solution. I had found that while animate and inanimate objects on Almuric often closely copied things on Earth, yet there was almost a striking difference somewhere, in substance, quality, shape or mode of action. It was preposterous that certain conditions on the separate planets could run such a perfect parallel as to produce an identical language. Yet I could not doubt the evidence of my ears. With a curse I abandoned the problem as too fantastic to waste time on.

Perhaps it was this incident, perhaps the glimpse of the distant savannas, which filled me with a restlessness and distaste for the barren hill country where I had fared so hardily. The sight of men, strange and alien as they were, stirred in my breast a desire for human companionship, and this frustrated longing became in turn a sudden feeling of repulsion for my surroundings. I did not hope to meet friendly humans on the plains; but I determined to try my chances upon them, nevertheless, though what perils I might meet there I could not know. Before I left the hills some whim caused me to scrape from my face my heavy growth and trim my shaggy hair with my poniard, which had lost none of its razor edge. Why I did this I cannot say, unless it was the natural instinct of a man setting forth into new country to look his "best."

The next morning I descended into the grassy plains, which swept eastward and southward as far as sight could reach. I continued eastward and covered many miles that day, without any unusual incident. I encountered several small winding rivers, along whose margins the grass stood taller than my head. Among this grass I heard the snorting and thrashing of heavy animals of some sort, and gave them a wide berth-for which caution I was later thankful.

The rivers were thronged in many cases with gaily colored birds of many shapes and hues, some silent, others continually giving forth strident cries as they wheeled above the waters or dipped down to snatch their prey from its depths.

Further out on the plain I came upon herds of grazing animals-small deer-like creatures, and a curious animal that looked like a pot-bellied pig with abnormally long hind legs, and that progressed in enormous bounds, after the fashion of a kangaroo. It was a most ludicrous sight, and I laughed until my belly ached. Later I reflected that it was the first time I had laughed-outside of a few short barks of savage satisfaction at the discomfiture of an enemy-since I had set foot on Almuric.

That night I slept in the tall grass not far from a water course, and might have been made the prey of any wandering meat-eater. But fortune was with me that night. All across the plains sounded the thunderous roaring of stalking monsters, but none came near my frail retreat. The night was warm and pleasant, strikingly in contrast with the nights in the chill grim hills.

The next day a momentous thing occurred. I had had no meat on Almuric, except when ravenous hunger had driven me to eat raw flesh. I had searched in vain for some stone that would strike a spark. The rocks were of a peculiar nature, unknown to Earth. But that morning on the plains, I found a bit of greenish-looking stone lying in the grass, and experiments showed that it had some of the qualities of flint. Patient effort, in which I clinked my poniard against the stone, rewarded me with a spark of fire in the dry grass, which I soon fanned to a blaze-and had some difficulty in extinguishing.

That night I surrounded myself with a ring of fire which I fed with dry grass and stalked plants which burned slowly and I felt comparatively safe, though huge forms moved about me in the darkness, and I caught the stealthy pad of great paws, and the glimmer of wicked eyes.

On my journey across the plains I subsisted on fruit I found growing on green stalks, which I saw the birds eating. It was pleasant to the taste, though lacking in the nutritive qualities of the nuts in the hills. I looked longingly at the scampering deer-like animals, now that I had the means of cooking their flesh, but saw no way of securing them.

And so for days I wandered aimlessly across those vast plains, until I came in sight of a massive walled city.

I sighted it just at nightfall, and eager though I was to investigate it further, I made my camp and waited for morning. I wondered if my fire would be seen by the inhabitants, and if they would send out a party to discover my nature and purpose.

With the fall of night I could no longer make it out, but the last waning light had shown it plainly, rising stark and somber against the eastern sky. At that distance no evidence of life was visible, but I had a dim impression of huge walls and massive towers, all of a greenish tint.

I lay within my circle of fire, while great sinuous bodies rustled through the grass and fierce eyes glared at me, and my imagination was at work as I strove to visualize the possible inhabitants of that mysterious city. Would they be of the same race as the hairy ferocious troglodytes I had encountered? I doubted it, for it hardly seemed possible that these primitive creatures would be capable of rearing such a structure. Perhaps there I would find a highly developed type of cultured man. Perhaps-here imaginings too dark and shadowy for description whispered at the back of my consciousness.

Then the moon rose behind the city, etching its massive outlines in the weird golden glow. It looked black and somber in the moonlight; there was something distinctly brutish and forbidding about its contours. As I sank into slumber I reflected that if ape-men could build a city, it would surely resemble that colossus in the moon.

CHAPTER 2

Dawn Found Me on my way across the plain. It may seem like the height of folly to have gone striding openly toward the city, which might be full of hostile beings, but I had learned to take desperate chances, and I was consumed with curiosity; weary at last of my lonely life.

The nearer I approached, the more rugged the details stood out. There was more of the fortress than the city about the walls, which, with the tower that loomed behind and above them, seemed to have been built of huge blocks of greenish stone, very roughly cut. There was no apparent attempt at smoothing, polishing, or otherwise adorning this stone. The whole appearance was rude and savage, suggesting a wild fierce people heaping up rocks as a defense against enemies.

As yet I had seen nothing of the inhabitants. The city might have been empty of human life. But a broad road leading to the massive gate was beaten bare of grass, as if by the constant impact of many feet. There were no fields or gardens about the city; the grass waved to the foot of the walls. All during that long march across the plain to the gates, I saw nothing resembling a human being. But as I came under the shadow of the great gate, which was flanked on either hand by a massive tower, I caught a glimpse of tousled black heads moving along the squat battlements. I halted and threw back my head to hail them. The sun had just topped the towers and its glare was full in my eyes. Even as I opened my lips, there was a cracking report like a rifle shot, a jet of white smoke spurted from a tower, and a terrific impact against my head dashed me into unconsciousness.

When I came to my senses it was not slowly, but quickly and clear-headedly, what with my immense recuperative powers. I was lying on a bare stone floor in a large chamber, the walls, ceiling and floor of which were composed of huge blocks of green stone. From a barred window high up in one wall sunlight poured to illuminate the room, which was without furnishing, except for a bench, crudely and massively built.

A heavy chain was looped about my waist and made fast with a strange, heavy lock. The other end of the chain was fastened to a thick ring set in the wall. Everything about the fantastic city seemed massive.

Lifting a hand to my head, I found it was bandaged with something that felt like silk. My head throbbed. Evidently whatever missile it was that had been fired at me from the wall, had only grazed my head, inflicting a scalp wound and knocking me senseless. I felt for my poniard, but naturally it was gone.

23

I cursed heartily. When I had found myself on Almuric I had been appalled by my prospects; but then at least I had been free. Now I was in the hands of God only knew what manner of beings. All I knew was that they were hostile. But my inordinate self-confidence would not down, and I felt no great fear. I did feel a rush of panic, common to all wild things, at being confined and shackled, but I fought down this feeling and it was succeeded by one of red unreasoning rage. Springing to my feet, which movement the chain was long enough to allow, I began jerking and tearing at my shackle.

It was while engaged in this fruitless exhibition of primitive resentment that a slight noise caused me to wheel, snarling, my muscles tensed for attack or defense. What I saw froze me in my tracks.

Just within the doorway stood a girl. Except in her garments she differed little from the type of girls I had known on Earth, except that her slim figure exhibited a suppleness superior to theirs. Her hair was intensely black, her skin white as alabaster. Her lissome limbs were barely concealed by a light, tunic-like garment, sleeveless, low-necked, revealing the greater part of her ivory breasts. This garment was girdled at her lithe waist, and came to within a few inches above her knees. Soft sandals encased her slender feet. She was standing in an attitude of awed fascination, her dark eyes wide, her crimson lips parted. As I wheeled and glared at her, she gave back with a quick gasp of surprise or fear, and fled lightly from the chamber.

I stared after her. If she were typical of the people of the city, then surely the effect produced by the brutish masonry was an illusion, for she seemed the product of some gentle and refined civilization, allowing for a certain barbaric suggestion about her costume.

While so musing, I heard the tramp of feet, harsh voices were lifted in argument, and the next instant a group of men strode into the chamber, halting as they saw me conscious and on my feet. Still thinking of the girl, I glared at them in surprise. They were of the same type as the others I had seen, huge, hairy, ferocious, with the same apelike forward-thrust heads and formidable faces. Some, I noticed, were darker than others, but all were dark and fierce, and the whole effect was one of somber and ferocious savagery. They were instinct with ferocity; it blazed in their icy-gray eyes, reflected in the snarling lift of their bristling lips, rumbled in their rough voices.

All were armed, and their hands seemed instinctively to seek their hilts as they stood glaring at me, their shaggy heads thrust forward in their apelike manner.

"Thak!" one exclaimed, or rather roared-all their voices were as gusty as a sea wind-"he's conscious!"

"Do you suppose he can speak or understand human language?" rumbled another.

All this while I had stood glaring back at them, wondering anew at their speech. Now I realized that they were not speaking English.

The thing was so unnatural that it gave me a shock. They were not speaking any Earthly language, and I realized it, yet I understood them, except for various words which apparently had no counterpart on Earth. I made no attempt to understand this seemingly impossible phenomenon, but answered the last speaker.

"I can speak and understand." I grunted. "Who are you? What city is this? Why did you attack me? Why am I in chains?"

They rumbled in amazement, with much tugging of mustaches, shaking of heads, and uncouth profanity.

"He talks, by Thak!" said one. "I tell you, he is from beyond the Girdle!"

"From beyond my hip!" broke in another rudely. "He is a freak, a damned, smooth-skinned degenerate misfit which should not have been born, or allowed to exist."

"Ask him how he came by the Bonecrusher's poniard," requested yet another.

"Did you steal this from Logar?" he demanded.

"I stole nothing!" I snapped, feeling like a wild beast being prodded through the bars of a cage by unfeeling and critical spectators. My rages, like all the emotions on that wild planet, were without restraint.

"I took that poniard from the man who carried it, and I took it in a fair fight," I added.

"Did you slay him?" they demanded unbelievingly.

"No," I growled. "We fought with our bare hands, until he tried to knife me. Then I knocked him senseless."

A roar greeted my words. I thought at first they were clamoring with rage;

25

then I made out that they were arguing among themselves.

"I tell you he lies!" one bull's bellow rose above the tumult. "We all know that Logar the Bonecrusher is not the man to be thrashed and stripped by a smooth-skinned hairless brown man like this. Ghor the Bear might be a match for Logar. No one else."

"Well, there's the poniard," someone pointed out.

The clamor rose again, and in an instant the disputants were yelling and cursing, and brandishing their hairy fists in one another's faces, hands fumbled at sword hilts, and challenges and defiances were exchanged freely.

I looked to see a general throat-cutting, but presently one who seemed in some authority drew his sword and began banging the hilt on the rude bench, at the same time drowning out the voices of the others with his bull-like bellowing.

"Shut up! Shut up! Let another man open his mouth and I'll split his head!" As the clamor subsided and the disputants glared venomously at him, he continued in a voice as calm as if nothing had occurred. "It's neither here nor there about the poniard. He might have caught Logar sleeping and brained him, or he might have stolen it, or found it. Are we Logar's brothers, that we should seek after his welfare?"

A general snarl answered this. Evidently the man called Logar was not popular among them.

"The question is, what shall we do with this creature? We've got to hold a council and decide. He's evidently uneatable." He grinned as he said this, which was apparently meant as a bit of grim humor.

"His hide would make good leather." suggested another in a tone that did not sound as though he was joking.

"Too soft," protested another.

"He didn't feel soft while we were carrying him in," returned the first speaker. "He was hard as steel springs."

"Tush," deprecated the other. "I'll show you how tender his flesh is. Watch me slice off a few strips." He drew his dagger and approached me while the others watched with interest.

26

All this time my rage had been growing until the chamber seemed to swim in a red mist. Now, as I realized that the fellow really intended trying the edge of his steel on my skin I went berserk. Wheeling, I gripped the chain with both hands, wrapping it around my wrists for more leverage. Then, bracing my feet against the floor and walls I began to strain with all my strength. All over my body the great muscles coiled and knotted; sweat broke out on my skin, and then with a shattering crash the stone gave way, the iron ring was torn out bodily, and I was catapulted on my back onto the floor, at the feet of my captors who roared with amazement and fell on me *en masse*.

I answered their bellows with one strident yell of blood-thirsty gratification, and heaving up through the melee, began swinging my heavy fists like caulking mallets. Oh, that was a rough-house while it lasted! They made no attempt to knife me, striving to swamp me with numbers. We rolled from one side of the chamber to the other, a gasping, thrashing, cursing, hammering mass, while with the yells, howls, earnest profanity, and impact of heavy bodies, it was a perfect bedlam. Once I seemed to catch a fleeting glimpse of the door thronged with the heads of women similar to the one I had seen, but I could not be sure; my teeth were set in a hairy ear, my eyes were full of sweat and stars from a vicious punch on the nose, and what with a gang of heavy forms romping all over me my sight was none too good.

Yet, I gave a good account of myself. Ears split, noses crumpled and teeth splintered under the crushing impact of my iron-hard fists, and the yells of the wounded were music to my battered ears. But that damnable chain about my waist kept tripping me and coiling about my legs, and pretty soon the bandage was ripped from my head, my scalp wound opened anew and deluged me with blood. Blinded by this I floundered and stumbled, and gasping and panting they bore me down and bound my arms and legs.

The survivors then fell away from me and lay or sat in positions of pain and exhaustion while I, finding my voice, cursed them luridly. I derived ferocious satisfaction at the sight of all the bloody noses, black eyes, torn ears and smashed teeth which were in evidence, and barked in vicious laughter when one announced with many curses that his arm was broken. One of them was out cold, and had to be revived, which they did by dumping over him a vessel of cold water that was fetched by someone I could not see from where I lay. I had no idea that it was a woman who came in answer to a harsh roar of command.

"His wound is open again," said one, pointing at me. "He'll bleed to death."

"I hope he does," snarled another, lying doubled up on the floor. "He's burst my belly. I'm dying. Get me some wine."

"If you're dying you don't need wine," brutally answered the one who seemed a chief, as he spat out bits of splintered teeth. "Tie up his wound, Akra."

Akra limped over to me with no great enthusiasm and bent down.

"Hold your damnable head still," he growled.

"Keep off!" I snarled. "I'll have nothing from you. Touch me at your peril."

He exasperatedly grabbed my face in his broad hand and shoved me violently down. That was a mistake. My jaws locked on his thumb, evoking an ear-splitting howl, and it was only with the aid of his comrades that he extricated the mangled member. Maddened by the pain, he howled wordlessly, then suddenly gave me a terrific kick in the temple, driving my wounded head with great violence back against the massive bench leg. Once again I lost consciousness.

When I came to myself again I was once more bandaged, shackled by the wrists and ankles, and made fast to a fresh ring, newly set in the stone, and apparently more firmly fixed than the other had been. It was night. Through the window I glimpsed the star-dotted sky. A torch which burned with a peculiar white flame was thrust into a niche in the wall, and a man sat on the bench, elbows on knees and chin on fists, regarding me intently. On the bench near him stood a huge gold vessel.

"I doubted if you'd come to after that last crack," he said at last.

"It would take more than that to finish me," I snarled. "You are a pack of cursed weaklings. But for my wound and that infernal chain, I'd have bested the whole mob of you."

My insults seemed to interest rather than anger him. He absently fingered a large bump on his head on which blood was thickly clotted, and asked: "Who are you? Whence do you come?"

"None of your business," I snapped.

He shrugged his shoulders, and lifting the vessel in one hand drew his dagger with the other.

"In Koth none goes hungry," he said, "I'm going to place this food near your hand and you can eat. But I warn you, if you try to strike or bite me, I'll stab you."

I merely snarled truculently, and he bent and set down the bowl, hastily withdrawing. I found the food to be a kind of stew, satisfying both thirst and hunger. Having eaten I felt in somewhat better mood, and my guard renewed his questions, I answered: "My name is Esau Cairn. I am an American, from the planet Earth."

He mulled over my statements for a space, then asked: "Are these places beyond the Girdle?"

"I don't understand you," I answered.

He shook his head. "Nor I you. But if you do not know of the Girdle, you cannot be from beyond it. Doubtless it is all fable, anyway. But whence did you come when we saw you approaching across the plain? Was that your fire we glimpsed from the towers last night?"

"I suppose so," I replied. "For many months I have lived in the hills to the west. It was only a few weeks ago that I descended into the plains."

He stared and stared at me.

"In the hills? Alone, and with only a poniard?"

"Well, what about it?" I demanded.

He shook his head as if in doubt or wonder. "A few hours ago I would have called you a liar. Now I am not sure."

"What is the name of this city?" I asked.

"Koth, of the Kothan tribe. Our chief is Khossuth Skull-splitter. I am Thab the Swift. I am detailed to guard you while the warriors hold council."

"What's the nature of their council?" I inquired.

"They discuss what shall be done with you; and they have been arguing since sunset, and are no nearer a solution than before."

"What is their disagreement?"

"Well," he answered. "Some want to hang you, and some want to shoot you."

29

"I don't suppose it's occurred to them that they might let me go," I suggested with some bitterness.

He gave me a cold look. "Don't be a fool," he said reprovingly.

At that moment a light step sounded outside, and the girl I had seen before tiptoed into the chamber. Thab eyed her disapprovingly.

"What are you doing here, Altha?" he demanded.

"I came to look again at the stranger," she answered in a soft musical voice. "I never saw a man like him. His skin is nearly as smooth as mine, and he has no hair on his countenance. How strange are his eyes! Whence does he come?"

"From the hills, he says," grunted Thab. Her eyes widened. "Why, none dwells in the hills, except wild beasts! Can it be that he is some sort of animal? They say he speaks and understands speech."

"So he does," growled Thab, fingering his bruises. "He also knocks out men's brains with his naked fists, which are harder and heavier than maces. Get away from there."

"He's a rampaging devil. If he gets his hands on you he won't leave enough of you for the vultures to pick."

"I won't get near him," she assured him. "But, Thab, he does not look so terrible. See, there is no anger in the gaze he fixes on me. What will be done with him?"

"The tribe will decide," he answered. "Probably let him fight a saber-tooth leopard bare-handed."

She clasped her own hands with more human feeling than I had yet seen shown on Almuric.

"Oh, Thab, why? He has done no harm; he came alone and with empty hands. The warriors shot him down without warning-and now-"

He glanced at her in irritation. "If I told your father you were pleading for a captive-"

Evidently the threat carried weight. She visibly wilted.

"Don't tell him," she pleaded. Then she flared up again. "Whatever you say, it's beastly! If my father whips me until the blood runs over my heels, I'll still say so!"

And so saying, she ran quickly out of the chamber.

"Who is that girl?" I asked.

"Altha, the daughter of Zal the Thrower."

"Who is he?"

"One of those you battled so viciously a short time ago."

"You mean to tell me a girl like that is the daughter of a man like-" Words failed me.

"What's wrong with her?" he demanded. "She differs none from the rest of our women."

"You mean all the women look like her, and all the men look like you?"

"Certainly-allowing for their individual characteristics. Is it otherwise among your people? That is, if you are not a solitary freak."

"Well, I'll be-" I began in bewilderment, when another warrior appeared in the door, saying.

"I'm to relieve you, Thab. The warriors have decide to leave the matter to Khossuth when he returns on the morrow."

Thab departed and the other seated himself on the bench. I made no attempt to talk to him. My head was swimming with the contradictory phenomena I had heard and observed, and I felt the need of sleep. I soon sank into dreamless slumber.

Doubtless my wits were still addled from the battering I had received. Otherwise I would have snapped awake when I felt something touch my hair. As it was, I woke only partly. From under drooping lids I glimpsed, as in a dream, a girlish face bent close to mine, dark eyes wide with frightened fascination, red lips parted. The fragrance of her foamy black hair was in my nostrils. She timidly touched my face, then drew back with a quick soft intake of breath, as if frightened by her action. The guard snored on the bench. The torch had burned to a stub that cast a weird dull glow over the chamber. Outside, the moon had set. This much I vaguely realized before I sank back into slumber again, to be haunted by a dim beautiful face that shimmered through my dreams.

CHAPTER 3

I Awoke Again in the cold gray light of dawn, at a time when the condemned meet their executioners. A group of men stood over me, and one I knew was Khossuth the Skullsplitter.

He was taller than most, and leaner-almost gaunt in comparison to the others. This circumstance made his broad shoulders seem abnormally huge. His face and body were seamed with old scars. He was very dark, and apparently old; an impressive and terrible image of somber savagery.

He stood looking down at me, fingering the hilt of his great sword. His gaze was gloomy and detached.

"They say you claim to have beaten Logar of Thurga in open fight," he said at last, and his voice was cavernous and ghostly in a manner I cannot describe.

I did not reply, but lay staring up at him, partly in fascination at his strange and menacing appearance, partly in the anger that seemed generally to be with me during those times.

"Why do you not answer?" he rumbled.

"Because I'm weary of being called a liar," I snarled.

"Why did you come to Koth?"

"Because I was tired of living alone among wild beasts. I was a fool. I thought I would find human beings whose company was preferable to the leopards and baboons. I find I was wrong."

He tugged his bristling mustaches.

"Men say you fight like a mad leopard. Thab says that you did not come to the gates as an enemy comes. I love brave men. But what can we do? If we free you, you will hate us because of what has passed, and your hate is not lightly to be loosed."

"Why not take me into the tribe?" I remarked, at random.

He shook his head. "We are not Yagas, to keep slaves."

33

"Nor am I a slave," I grunted. "Let me live among you as an equal. I will hunt and fight with you. I am as good a man as any of your warriors."

At this another pushed past Khossuth. This fellow was bigger than any I had yet seen in Koth-not taller, but broader, more massive. His hair was thicker on his limbs, and of a peculiar rusty cast instead of black.

"That you must prove!" he roared, with an oath. "Loose him, Khossuth! The warriors have been praising his power until my belly revolts! Loose him and let us have a grapple!"

"The man is wounded, Ghor," answered Khossuth.

"Then let him be cared for until his wound is healed," urged the warrior eagerly, spreading his arms in a curious grappling gesture.

"His fists are like hammers," warned another.

"Thak's devils!" roared Ghor, his eyes glaring, his hairy arms brandished. "Admit him into the tribe, Khossuth! Let him endure the test! If he survives-well, by Thak, he'll be worthy even to be called a man of Koth!"

"I will go and think upon the matter," answered Khossuth after a long deliberation.

That settled the matter for the time being. All trooped out after him. Thab was last, and at the door he turned and made a gesture which I took to be one of encouragement. These strange people seemed not entirely without feelings of pity and friendship.

The day passed uneventfully. Thab did not return. Other warriors brought me food and drink, and I allowed them to bandage my scalp. With more human treatment the wild-beast fury in me had been subordinated to my human reason. But that fury lurked close to the surface of my soul, ready to blaze into ferocious life at the slightest encroachment.

I did not see the girl Altha, though I heard light footsteps outside the chamber several times, whether hers or another's I could not know.

About nightfall a group of warriors came into the room and announced that I was to be taken to the council, where Khossuth would listen to all arguments and decide my fate. I was surprised to learn that arguments would be presented on my behalf. They got my promise not to attack them, and loosed me from the chain that bound me to the wall, but they did not

34

remove the shackles on my wrists and ankles.

I was escorted out of the chamber into a vast hall, lighted by white fire torches. There were no hangings or furnishings, nor any sort of ornamentation-just an almost oppressive sense of massive architecture.

We traversed several halls, all equally huge and windy, with rugged walls and lofty ceilings, and came at last into a vast circular space, roofed with a dome. Against the back wall a stone throne stood on a block-like dais, and on the throne sat old Khossuth in gloomy majesty, clad in a spotted leopardskin. Before him in a vast three-quarters circle sat the tribe, the men cross-legged on skins spread on the stone floor, and behind them the women and children seated on fur-covered benches.

It was a strange concourse. The contrast was startling between the hairy males and the slender, white-skinned, dainty women. The men were clad in loincloths and high-strapped sandals; some had thrown panther-skins over their massive shoulders. The women were dressed similar to the girl Altha, whom I saw sitting with the others. They wore soft sandals or none, and scanty tunics girdled about their waists. That was all. The difference of the sexes was carried out down to the smallest babies. The girl children were quiet, dainty and pretty. The young males looked even more like monkeys than did their elders.

I was told to take my seat on a block of stone in front and somewhat to the side of the dais. Sitting among the warriors I saw Ghor, squirming impatiently as he unconsciously flexed his thick biceps.

As soon as I had taken my seat, the proceedings went forward. Khossuth simply announced that he would hear the arguments, and pointed out a man to represent me, at which I was again surprised, but this apparently was a regular custom among these people. The man chosen was the lesser chief who had commanded the warriors I had battled in the cell, and they called him Gutchluk Tigerwrath. He eyed me venemously as he limped forward with no great enthusiasm, bearing the marks of our encounter.

He laid his sword and dagger on the dais, and the foremost warriors did likewise. Then he glared at the rest truculently, and Khossuth called for arguments to show why Esau Cairn-he made a marvelous jumble of the pronunciation-should not be taken into the tribe.

Apparently the reasons were legion. Half a dozen warriors sprang up and began shouting at the top of their voice, while Gutchluk dutifully strove to answer them. I felt already doomed. But the game was not played out, or even well begun. At first Gutchluk went at it only half-heartedly, but opposition heated him to his talk. His eyes blazed, his jaw jutted, and he began to roar and bellow with the best of them. From the arguments he presented, or rather thundered, one would have thought he and I were lifelong friends.

No particular person was designated to protest against me. Everybody who wished took a hand. And if Gutchluk won over anyone, that person joined his voice to Gutchluk's. Already there were men on my side. Thab's shout and Ghor's bellow vied with my attorney's roar, and soon others took up my defense.

That debate is impossible for an Earthman to conceive of, without having witnessed it. It was sheer bedlam, with from three voices to five hundred voices clamoring at once. How Khossuth sifted any sense out of it, I cannot even guess. But he brooded somberly above the tumult, like a grim god over the paltry aspirations of mankind.

There was wisdom in the discarding of weapons. Dispute frequently became biting, and criticisms of ancestors and personal habits entered into it. Hands clutched at empty belts and mustaches bristled belligerently. Occasionally Khossuth lifted his weird voice across the clamor and restored a semblance of order.

My attempts to follow the arguments were vain. My opponents went into matters seemingly utterly irrelevant, and were met by rebuttals just as illogical. Authorities of antiquity were dragged out, to be refuted by records equally musty.

To further complicate matters, disputants frequently snared themselves in their own arguments, or forgot which side they were on, and found themselves raging frenziedly on the other. There seemed no end to the debate, and no limit to the endurance of the debaters. At midnight they were still yelling as loudly, and shaking their fists in one another's beards as violently as ever.

The women took no part in the arguments.

They began to glide away about midnight, with the children. Finally only one small figure was left among the benches. It was Altha, who was following-or trying to follow-the proceedings with a surprising interest.

I had long since given up the attempt. Gutchluk was holding the floor valiantly, his veins swelling and his hair and beard bristling with his exertions. Ghor was actually weeping with rage and begging Khossuth to let him break a few necks. Oh, that he had lived to see the men of Koth become adders and snakes, with the hearts of buzzards and the guts of toads! he bawled, brandishing his huge arms to high heaven.

It was all a senseless madhouse to me. Finally, in spite of the clamor, and the fact that my life was being weighed in the balance, I fell asleep on my block and snored peacefully while the men of Koth raged and pounded their hairy breasts and bellowed, and the strange planet of Almuric whirled on its way under the stars that neither knew nor cared for men, Earthly or otherwise.

It was dawn when Thab shook me awake and shouted in my ear: "We have won! You enter the tribe, if you'll wrestle Ghor!"

"I'll break his back!" I grunted, and went back to sleep again.

CHAPTER 4

So began my life as a man among men on Almuric. I who had begun my new life as a naked savage, now took the next step on the ladder of evolution and became a barbarian. For the men of Koth were barbarians, for all their silks and steel and stone towers. Their counterpart is not on Earth today, nor has it ever been. But of that later. Let me tell first of my battle with Ghor the Bear.

My chains were removed and I was taken to a stone tower on the wall, there to dwell until my wounds had healed. I was still a prisoner. Food and drink were brought me regularly by the tribesmen, who also tended carefully to my wounds, which were unimportant, considering the hurts I had had from wild beasts, and had recovered from unaided. But they wished me to be in prime condition for the wrestling, which was to decide whether I should be admitted to the tribe of Koth, or-well, from what they said of Ghor, if I lost there would be no problem as to my disposition. The wolves and vultures would take care of that.

Their manner toward me was noncommittal, with the exception of Thab the Swift, who was frankly cordial to me. I saw neither Khossuth, Ghor nor Gutchluk during the time I was imprisoned in the tower, nor did I see the girl Altha.

I do not remember a more tedious and wearisome time. I was not nervous because of any fear of Ghor; I frankly doubted my ability to beat him, but I had risked my life so often and against such fearful odds, that personal fear had been stamped out of my soul. But for months I had lived like a mountain panther, and now to be caged up in a stone tower, where my movements were limited, bounded and restricted-it was intolerable, and if I had been forced to put up with it a day longer, I would have lost control of myself, and either fought my way to freedom or perished in the attempt. As it was, all the constrained energy in me was pent up almost to the snapping point, giving me a terrific store of nervous power which stood me in good stead in my battle.

There is no man on Earth equal in sheer strength to any man of Koth. They lived barbaric lives, filled with continuous peril and warfare against foes human and bestial. But after all, they lived the lives of men, and I had been living the life of a wild beast.

As I paced my tower chamber, I thought of a certain great wrestling champion of Europe with whom I had once contested in a friendly private bout, and who pronounced me the strongest man he had ever encountered.

Could he have seen me now, in the tower of Koth! I am certain that I could have torn out his biceps like rotten cloth, broken his spine across my knee, or caved in his breastbone with my clenched fist; and as for speed, the most finely trained Earth athlete would have seemed awkward and sluggish in comparison to the tigerish quickness lurking in my rippling sinews.

Yet for all that, I knew that I would be strained to the uttermost even to hold my own with the giant they called Ghor the Bear. He did, indeed, resemble a shaggy rusty-hued cave-bear.

Thab the Swift narrated some of his triumphs to me, and such a record of personal mayhem I never heard; the man's progress through life was marked by broken limbs, backs and necks. No man had yet stood before him in barehanded battle, though some swore Logar the Bonecrusher was his equal.

Logar, I learned, was chief of Thugra, a city hostile to Koth. All cities on Almuric seemed to be hostile to each other, the people of the planet being divided into many small tribes, incessantly at war. The chief of Thugra was called the Bonecrusher because of his terrible strength. The poniard I had taken from him had been his favorite weapon, a famous blade, forged, Thab said, by a supernatural smith. Thab called this being a *gorka*, and I found in tales concerning the creature an analogy to the dwarfish metalworkers of the ancient Germanic myths of my own world.

Thab told me much concerning his people and his planet, but of these things I will deal later. At last Khossuth came, found my wounds completely cured, eyed my bronzed sinews with a shadow of respect in his cold brooding eyes, and pronounced me fit for battle.

Night had fallen when I was led into the streets of Koth. I looked with wonder at the giant walls towering above me, dwarfing their human inhabitants. Everything in Koth was built on a heroic scale. Neither the walls nor the edifices were unusually high, in comparison to their bulk, but they were so massive. My guides led me to a sort of amphitheater near the outer wall. It was an oval space surrounded by huge stone blocks, rising tier upon tier, and forming seats for the spectators. The open space in the center was hard ground, covered with short grass. A sort of bulwark was formed about it out of woven leather thongs, apparently to keep the contestants from dashing their heads against the surrounding stones. Torches lighted the scene.

The spectators were already there, the men occupying the lower blocks, the women and children the upper. My gaze roved over the sea of faces, hairy or smooth, until it rested on one I recognized, and I felt a strange throb of pleasure at the sight of Altha sitting there watching me with her intent dark

39

eyes.

Thab indicated for me to enter the arena, and I did so, thinking of the old-time bare-knuckled bouts of my own planet, which were fought in crude rings pitched, like this, on the naked turf. Thab and the other warriors who had escorted me remained outside. Above us brooded old Khossuth on a carven stone elevated above the first tier, and covered with leopard-skins.

I glanced beyond him to that dusky star-filled sky whose strange beauty never failed to fascinate me, and I laughed at the fantasy of it all-where I, Esau Cairn, was to earn by sweat and blood my right to exist on this alien world, the existence of which was undreamed by the people of my own planet.

I saw a group of warriors approaching from the other side, a giant form looming among them. Ghor the Bear glared at me across the ring, his hairy paws grasping the thongs, then with a roar he vaulted over them and stood before me, an image of truculence incarnate-angry because I had chanced to reach the ring before him.

On his rude throne above us, old Khossuth lifted a spear and cast it earthward. Our eyes followed its flight, and as it sheathed its shining blade in the turf outside the ring, we hurled ourselves at each other, iron masses of bone and thew, vibrant with fierce life and the lust to destroy.

We were each naked except for a sort of leather loin-clout, which was more brace than garment. The rules of the match were simple, we were not to strike with our fists or open hands, knees or elbows, kick, bite or gouge. Outside of that, anything went.

At the first impact of his hairy body against mine, I realized that Ghor was stronger than Logar. Without my best natural weapons-my fists-Ghor had the advantage.

He was a hairy mountain of iron muscle, and he moved with the quickness of a huge cat. Accustomed to such fighting, he knew tricks of which I was ignorant. Lastly, his bullet head was set so squarely on his shoulders that it was practically impossible to strangle that thick squat neck of his.

What saved me was the wild life I had lived which had toughened me as no man, living as a man, can be toughened. Mine was the superior quickness, and ultimately, the superior endurance.

There is little to be said of that fight. Time ceased to be composed of intervals of change, and merged into a blind mist of tearing, snarling

40

eternity. There was no sound except our panting gasps, the guttering of the torches in the light wind, and the impact of our feet on the turf, of our hard bodies against each other. We were too evenly matched for either to gain a quick advantage. There was no pinning of shoulders, as in an Earthly wrestling match. The fight would continue until one or both of the contestants were dead or senseless.

When I think of our endurance and stamina, I stand appalled. At midnight we were still rending and tearing at each other. The whole world was swimming red when I broke free out of a murderous grapple. My whole frame was a throb of wrenched, twisted agony. Some of my muscles were numbed and useless. Blood poured from my nose and mouth. I was half blind and dizzy from the impact of my head against the hard earth. My legs trembled and my breath came in great gulps. But I saw that Ghor was in no better case. He too bled at the nose and mouth, and more, blood trickled from his ears. He reeled as he faced me, and his hairy chest heaved spasmodically. He spat out a mouthful of blood, and with a roar that was more a gasp, he hurled himself at me again. And steeling my ebbing strength for one last effort, I caught his outstretched wrist, wheeled, ducking low and bringing his arm over my shoulder, and heaved with all my last ounce of power.

The impetus of his rush helped my throw. He whirled headlong over my back and crashed to the turf on his neck and shoulder, slumped over and lay still. An instant I stood swaying above him, while a sudden deep-throated roar rose from the people of Koth, and then a rush of darkness blotted out the stars and the flickering torches, and I fell senseless across the still body of my antagonist.

Later they told me that they thought both Ghor and I were dead. They worked over us for hours. How our hearts resisted the terrible strain of our exertions is a matter of wonder to me. Men said it was by far the longest fight ever waged in the arena.

Ghor was badly hurt, even for a Kothan. That last fall had broken his shoulder bone and fractured his skull, to say nothing of the minor injuries he had received before the climax. Three of my ribs were broken, and my joints, limbs and muscles so twisted and wrenched that for days I was unable even to rise from my couch. The men of Koth treated our wounds and bruises with all their skill, which far transcends that of the Earth; but in the main it was our remarkable primitive vitality that put us back on our feet. When a creature of the wild is wounded, he generally either dies quickly or recovers quickly.

I asked Thab if Ghor would hate me for his defeat, and Thab was at a loss; Ghor had never been defeated before.

41

But my mind was soon put to rest on this score. Seven brawny warriors entered the chamber in which I had been placed, bearing a litter on which lay my late foe, wrapped in so many bandages he was scarcely recognizable. But his bellowing voice was familiar. He had forced his friends to bring him to see me as soon as he was able to stir on his couch. He held no malice. In his great, simple, primitive heart there was only admiration for the man who had given him his first defeat. He recounted our Homeric struggle with a gusto that made the roof reverberate, and roared his impatient eagerness for us to fare forth and do battle together against the foes of Koth.

He was borne back to his own chamber, still bellowing his admiration and gory plans for the future, and I experienced a warm glow in my heart for this great-hearted child of nature, who was far more of a man than many sophisticated scions of civilization that I had met.

And so I, Esau Cairn, took the step from savagery to barbarism. In the vast domed council hall before the assembled tribesmen, as soon as I was able, I stood before the throne of Khossuth Skullsplitter, and he cut the mysterious symbol of Koth above my head with his sword. Then with his own hands he buckled on me the harness of a warrior-the broad leather belt with the iron buckle, supporting my poniard and a long straight sword with a broad silver guard. Then the warriors filed past me, and each chief placed his palm against mine, and spoke his name, and I repeated it, and he repeated the name they had given me: Ironhand. That part was most wearisome for there were some four thousand warriors, and four hundred of these were chiefs of various rank. But it was part of the ritual of initiation, and when it was over I was as much a Kothan as if I had been born into the tribe.

In the tower chamber, pacing like a caged tiger while Thab talked, and later as a member of the tribe, I learned all that the people of Koth knew of their strange planet.

They and their kind, they said, were the only true humans on Almuric, though there was a mysterious race of beings dwelling far to the south called Yagas. The Kothans called themselves Guras, which applied to all cast in their mold, and meant no more than "man" does on Earth. There were many tribes of Guras, each dwelling in its separate city, each of which was a counterpart of Koth. No tribe numbered more than four or five thousand fighting-men, with the appropriate number of women and children.

No man of Koth had ever circled the globe, but they ranged far in their hunts and raids, and legends had been handed down concerning their world-which, naturally, they called by a name simply corresponding to the word "Earth"; though after a while some of them took up my habit of speaking of the planet as Almuric. Far to the north there was a land of ice and snow, uninhabited by human beings, though men spoke of weird cries shuddering by night from the ice crags, and of shadows falling across the snow. A lesser distance to the south rose a barrier no man had ever passed- a gigantic wall of rock which legend said girdled the planet; it was called, therefore, the Girdle. What lay beyond that Girdle, none knew. Some believed it was the rim of the world, and beyond it lay only empty space. Others maintained that another hemisphere lay beyond it. They believed, as seemed to me most logical, that the Girdle separated the northern and southern halves of the world, and that the southern hemisphere was inhabited by men and animals, though the exponents of their theory could give no proof, and were generally scoffed at as over-imaginative romanticists.

At any rate, the cities of the Guras dotted the vast expanse that lay between the Girdle and the land of ice. The northern hemisphere possessed no great body of water. There were rivers, great plains, a few scattered lakes, occasional stretches of dark, thick forests, long ranges of barren hills, and a few mountains. The larger rivers ran southward, to plunge into chasms in the Girdle.

The cities of the Guras were invariably built on the open plains, and always far apart. Their architecture was the result of the peculiar evolution of their builders-they were, basically, fortresses of rocks heaped up for defense. They reflected the nature of their builders, being rude, stalwart, massive, despising gaudy show and ornamentation, and knowing nothing of the arts.

In many ways the Guras are like the men of Earth, in other ways bafflingly different. Some of the lines on which they have evolved are so alien to Earthly evolution that I find it difficult to explain their ways and their development.

Specifically, Koth-and what is said of Koth can be said of every other Gura city:-the men of Koth are, skilled in war, the hunt, and weapon-making. The latter science is taught to each male child, but now seldom used. It is seldom found necessary to manufacture new arms, because of the durability of the material used. Weapons are handed down from generation to generation, or captured from enemies.

Metal is used only for weapons, in building, and for clasps and buckles on garments. No ornaments are worn, either by men or women, and there are no such things as coins. There is no medium of exchange. No trade between cities exists, and such "business" as goes on within the city is a matter of barter. The only cloth worn is a kind of silk, made from the fiber of a curious plant grown within the city walls. Other plants furnish wine, fruit, and seasonings. Fresh meat, the principal food of the Guras, is furnished by hunting, a pastime at once a sport and an occupation.

The folk of Koth, then, are highly skilled in metal-working, in silk-weaving, and in their peculiar form of agriculture. They have a written language, a simple hieroglyphic form, scrawled on leaves like papyrus, with a dagger-like pen dipped in the crimson juice of a curious blossom, but few except the chiefs can read or write. Literature they have none; they know nothing of painting, sculpturing, or the "higher" learning. They have evolved to the point of culture needful for the necessities of life, and they progress no further. Seemingly defying laws we on Earth have come to regard as immutable, they remain stationary, neither advancing nor retrogressing.

Like most barbaric people, they have a form of rude poetry, dealing almost exclusively with battle, mayhem and rapine. They have no bards or minstrels, but every man of the tribe knows the popular ballads of his clan, and after a few jacks of ale is prone to bellow them forth in a voice fit to burst one's eardrums.

These songs are never written down, and there is no written history. As a result, events of antiquity are hazy, and mixed with improbable legends.

No one knows how old is the city of Koth. Its gigantic stones are impervious to the elements, and might have stood there ten years or ten thousand years. I am of the opinion that the city is at least fifteen thousand years old. The Guras are an ancient race, in spite of their exuberant barbarism which gives them the atmosphere of a new young people. Of the evolution of the race from whatever beast was their common ancestor, of their racial splittings off and tribal drifts, of their development to their present condition, nothing whatever is known. The Guras themselves have no idea of evolution. They suppose that, like eternity, their race is without beginning and without end, that they have always been exactly as they are now. They have no legends to explain their creation.

I have devoted most of my remarks to the men of Koth. The women of Koth are no less worthy of detailed comment. I found the difference in the appearance of the sexes not so inexplicable after all. It is simply the result of natural evolution, and its roots lie in a fierce tenderness on the part of the Gura males for their women. It was to protect their women that they first, I am certain, built those brutish heaps of stone and dwelt among them; for the innate nature of the Gura male is definitely nomadic.

The woman, carefully guarded and shielded both from danger and from the hard work that is the natural portion of the women of Earthly barbarians, evolved by natural process into the type I have described. The men, on the other hand, lead incredibly active and strenuous lives. Their existence has been a savage battle for survival, ever since the first ape stood upright on Almuric. And they have evolved into a special type to fit their needs. Their peculiar appearance is not a result of degeneration or underdevelopment. They are, indeed, a highly specialized type, finely adapted to the wild life they follow.

As the men assume all risks and responsibility, they naturally assume all authority. The Gura woman has no say whatever in the government of the city and tribe, and her mate's authority over her is absolute, with the exception that she has the right to appeal to the council and chief in case of oppression. Her scope is narrow; few women ever set foot outside the city in which they are born, unless they are carried off in a raid.

Yet her lot is not so unhappy as it might seem. I have said that one of the characteristics of the Gura male is a savage tenderness for his women. Mistreatment of a woman is very rare, not tolerated by the tribe.

Monogamy is the rule. The Guras are not given to hand-kissing and pretty compliments, and the other superficial adjuncts of chivalry, but there is justice and a rough kindness in their dealing with women, somewhat similar to the attitude of the American frontiersman.

The duties of the Gura women are few, concerned mainly with child-bearing and child-rearing. They do no work heavier than the manufacturing of silk from the silk plants. They are musically inclined, and play on a small, stringed affair, resembling a lute, and they sing. They are quicker-witted, and of much more sensitive mind than the men. They are witty, merry, affectionate, playful and docile. They have their own amusements, and time does not seem to drag for them. The average woman could not be persuaded to set foot outside the city walls. They well know the perils that hem the cities in, and they are content in the protection of their ferocious mates and masters.

The men are, as I have said, in many ways like barbaric peoples on Earth. In some respects they resemble, I imagine, the ancient Vikings. They are honest, scorning theft and deceit. They delight in war and the hunt, but are not wantonly cruel, except when maddened by rage or bloodlust. Then they can be screaming fiends. They are blunt in speech, rough in their manners, easily angered, but as easily pacified, except when confronted by an hereditary enemy. They have a definite, though crude, sense of humor, a ferocious love for tribe and city, and a passion for personal freedom.

Their weapons consist of swords, daggers, spears, and a firearm something like a carbine-a single-shot, breech-loading weapon of no great range. The combustible material is not powder, as we know it. Its counterpart is not found on Earth. It possesses both percussion and explosive qualities. The bullet is of a substance much like lead. These weapons were used mainly in war with men; for hunting, bows and arrows were most often used.

Hunting parties are always going forth, so that the full force of warriors is seldom in the city at once. Hunters are often gone for weeks or months. But there are always a thousand fighting men in the city to repel possible attack, though it is not often that the Guras lay siege to a hostile city. Those cities are difficult to storm, and it is impossible to starve out the inhabitants, since they produce so much of their food supply within the walls, and in each city is an unfailing spring of pure water. The hunters frequently sought their prey in the hills which I had haunted, and which were reputed to contain more and varied forms of ferocious bestial life than any other section of the globe. The boldest hunters went in strong parties to the ills, and seldom roamed there more than a few days. The fact that I had lived among the hills alone for months won me even more respect and admiration among those wild fighting men than had my fight with Ghor.

Oh, I learned much of Almuric. As this is a chronicle and not an essay, I can scarcely skim the surface of customs, ways and traditions. I learned all they could tell me, and I learned much more. The Guras were not first on Almuric, though they considered themselves to be. They told me of ancient ruins, never built by Guras, relics of vanished races, who, they supposed, were contemporary with their distant ancestors, but which, as I came to learn, had risen and vanished awfully before the first Gura began to heap up stones to build his primordial city. And how I learned what no Gura knew is part of this strange narrative.

But they spoke of strange inhuman beings or survivals. They told me of the Yagas, a terrible race of winged black men, dwelling far to the south, within sight of the Girdle, in the grim city of Yugga, on the rock Yuthla, by the River Yogh, in the land of Yagg, where living man had never set foot. The Yagas, the Guras said, were not true men, but devils in a human form. From Yugga they swooped periodically, bearing the sword of slaughter and the torch of destruction, to carry young Gura girls into a slavery the manner of which none knew, because none had ever escaped from the land of Yagg. Some men thought that they were fed to a monster worshiped by the Yagas as a god, though some swore that the fiends worshiped nothing except themselves. This was known: their ruler was a black queen, named Yasmeena, and for more than a thousand years she had reigned on the grim rock of Yuthla, her shadow falling across the world to make men shudder.

The Guras told me other things, things weird and terrible: of dog-headed monstrosities skulking beneath the ruins of nameless cities; of earth-shaking colossals stalking through the night; of fires flitting like flaming bats through the shadowy skies; of things that haunted midnight forests,

46

crawling, squamous things that were never seen, but which tracked men down in the dank depths. They told me of great bats whose laughter drove men mad, and of gaunt shapes shambling hideously through the dusk of the hills. They told me of such things as do not even haunt the dreams of men on my native planet. For Life has taken strange shapes on Almuric, and natural Life is not the only Life there.

But the nightmares told to me and the nightmares seen by me unfold in their place, and I have already lingered too long in my narrative. Be patient a little, because events move swiftly on Almuric, and my chronicle moves no less swiftly when well under way.

For months I dwelt in Koth, fitting into the life of hunting, feasting, ale-guzzling, and brawling, as if I had been born into it. There life was not restrained and bound down, as it is on Earth. As yet no tribal war had tested my powers, but there was fighting enough in the city with naked hands, in friendly bouts, and drunken brawls, when the fighting-men dashed down their foaming jacks and bellowed their challenges across the ale-stained boards. I reveled in my new existence. Here, as in the hills, I threw my full powers unleashed into life; and here, unlike as in the hills, I had human companionship, of a sort that suited my particular make-up. I felt no need of art, literature or intellectuality; I hunted, I gorged, I guzzled, I fought; I spread my massive arms and clutched at life like a glutton. And in my brawling and reveling I all but forgot the slender figure which had sat so patiently in the council chamber beneath the great dome.

CHAPTER 5

I had wandered far in my hunting. Alone I had spent several nights on the plains. Now I was returning leisurely, but I was still many miles from Koth, whose massive towers I could not yet glimpse across the waving savannas. I cannot say what my thoughts were as I swung along, my carbine in the crook of my arm, but they were likely concerned with spoors in the water's edge, crushed-down grass marking the passing of some large animal, or the scents borne on the light wind.

Whatever my thoughts may have been, they were interrupted by a shrill cry. Wheeling, I saw a slim white figure racing across the grassy level toward me. Behind her, gaining with every stride, came one of those giant carnivorous birds which are among the most dangerous of all the grisly denizens of the grasslands. They tower ten feet in height and somewhat resemble an ostrich except for the beak, which is a huge curving weapon, three feet in length, pointed and edged like a scimitar. A stroke of that beak can slash a man asunder, and the great taloned feet of the monster can tear a human limb from limb.

This mountain of destruction was hurtling along behind the flying girl at appalling speed, and I knew it would overtake her long before I could hope to reach them. Cursing the necessity for depending on my none too accurate marksmanship, I lifted my carbine and took as steady an aim as possible. The girl was directly in line with the brute, and I could not risk a shot at the huge body, lest I hit her instead. I had to chance a shot at the great head that bobbed bafflingly on the long arching neck.

It was more luck than skill that sent my bullet home. At the crack of the shot the giant head jerked backward as if the monster had run into an unseen wall. The stumpy wings thrashed thunderously, and staggering erratically, the brute pitched to the earth.

The girl fell at the same instant, as if the same bullet had brought them both down. Running forward to bend over her, I was surprised to see Altha, daughter of Zal, looking up at me with her dark enigmatic eyes. Quickly satisfying myself that she was not injured, outside of fright and exhaustion, I turned to the thunderbird and found it quite dead, its few brains oozing out of a hole in its narrow skull.

Turning back to Altha, I scowled down at her.

"What are you doing outside the city?" I demanded. "Are you quite mad, to venture so far into the wilderness alone?"

She made no reply, but I sensed a hurt in her dark eyes, and I repented the roughness of my speech. I dropped down on one knee beside her.

"You are a strange girl, Altha," I said. "You are not like the other women of Koth. Folk say you are willful and rebellious, without reason. I do not understand you. Why should you risk your life like this?"

"What will you do now?" she demanded.

"Why, take you back to the city, of course."

Her eyes smoldered with a curious sullenness.

"You will take me back, and my father will whip me. But I will run away again-and again-and again!"

"But why should you run away?" I asked in bewilderment. "There is nowhere for you to go. Some beast will devour you."

"So!" she answered. "Perhaps it is my wish to be devoured."

"Then why did you run from the thunderbird?"

"The instinct to live is hard to conquer," she admitted.

"But why should you wish to die?" I expostulated. "The women of Koth are happy, and you have as much as any."

She looked away from me, out across the broad plains.

"To eat, drink and sleep is not all," she answered in a strange voice. "The beasts do that."

I ran my fingers through my thick hair in perplexity. I had hard similar sentiments voiced in many different ways on Earth, but it was the first time I had ever heard them from the lips of an inhabitant on Almuric. Altha continued in a low detached voice, almost as if she were speaking to herself rather than to me:

"Life is too hard for me. I do not fit, somehow, as the others do. I bruise myself on its rough edges. I look for something that is not and never was."

Uneasy at her strange words, I caught her heavy locks in my hands and forced back her head to look into her face. Her enigmatic eyes met mine with a strange glimmer in them such as I had never seen.

"It was hard before you came," she said. "It is harder now."

Startled, I released her, and she turned her head away.

"Why should I make it harder?" I asked bewilderedly.

"What constitutes life?" she countered. "Is the life we live all there is? Is there nothing outside and beyond our material aspirations?"

I scratched my head in added perplexity.

"Why," I said, "on Earth I met many people who were always following some nebulous dream or ideal, but I never observed that they were happy. On my planet there is much grasping and groping for unseen things, but I never knew there was such full content as I have known on Almuric."

"I thought you different," she said, still looking away from me. "When I saw you lying wounded and in chains, with your smooth skin and strange eyes, I thought you were gentler than other men. But you are as rough and fierce as the rest. You spend your days and nights in slaying beasts, fighting men, and in riotous wassail."

"But they all do," I protested.

She nodded. "And so I do not fit in life, and were better dead."

I felt unreasonably ashamed. It had occurred to me that an Earth-woman would find life on Almuric intolerably crude and narrow, but it seemed beyond reason that a native woman would have such feelings. If the other women I had seen desired more superficial gentleness on the part of their men, they had not made it known. They seemed content with shelter and protection, and cheerfully resigned to the rough manners of the males. I sought for words but found none, unskilled as I was in polite discourse. I suddenly felt my roughness, crudity and raw barbarism, and stood abashed.

"I'll take you back to Koth," I said helplessly.

She shrugged her shapely shoulders. "And you can watch my father whip me, if you will."

At that I found my tongue.

50

"He won't whip you," I retorted angrily. "Let him lay a hand on you, and I'll break his back."

She looked up at me quickly, with eyes widened in sudden interest. My arm had found its way about her slim form, and I was glaring into her eyes, with my face very close to hers. Her lips parted, and had that breathless instant lasted a little longer, I know not what would have happened. But suddenly the color went from her face, and from her parted lips rang a terrible scream. Her gaze was fixed on something beyond and above me, and the thrash of wings suddenly filled the air.

I wheeled on one knee, to see the air above me thronged with dark shapes. The Yagas! The winged men of Almuric! I had half believed them a myth; yet here they were in all their mysterious terror.

I had but a glance as I reared up, clubbing my empty carbine. I saw that they were tall and rangy in build, sinewy and powerful, with ebon skins. They seemed made like ordinary men, except for the great leathery bat-like wings which grew from their shoulders. They were naked except for loincloths, and were armed with short curved blades.

I rose on my toes as the first swooped in, scimitar lifted, and met him with a swing of my carbine that broke off the stock and crushed his narrow skull like an eggshell. The next instant they were whirling and thrashing about me, their curved blades licking at me like jets of lightning from all sides, the very number of their broad wings hampering them.

Whirling the carbine barrel in a wheel about me, I broke and beat back the flickering blades, and in a furious exchange of strokes, caught another a glancing blow on the head that stretched him senseless at my feet. Then a wild despairing cry rang out behind me, and abruptly the rush slackened.

The whole pack was in the air, racing southward, and I stood frozen. In the arms of one of them writhed and shrieked a slender white figure, stretching out imploring arms to me. Altha! They had snatched her up from behind my back, and were carrying her away to whatever doom awaited her in that black citadel of mystery far to the south. The terrific velocity with which the Yagas raced through the sky was already taking them out of my sight.

As I stood there baffled, I felt a movement at my feet. Looking down I saw one of my victims sit up and feel his head dazedly. I vengefully lifted my carbine barrel to dash out his brains; then a sudden thought struck me, inspired by the ease with which Altha's captor had carried both his weight and hers in the air.

51

Drawing my poniard, I dragged my captive to his feet. Standing erect he was taller than I, with shoulders equally broad, though his limbs were lean and wiry rather than massive. His dark eyes, which slanted slightly, regarded me with the unblinking stare of a venomous serpent.

The Guras had told me the Yagas spoke a tongue similar to their own.

"You are going to carry me through the air in pursuit of your companions," I said.

He shrugged his shoulders and spoke in a peculiarly harsh voice.

"I cannot carry your weight."

"Then that's too bad for you," I answered grimly, and whirling him about, I leaped upon his back, locking my legs about his waist. My left arm was hooked about his neck, the poniard in my right hand pricked his side. He had kept his feet under the impact of my bulk, spreading his great wings.

"Take the air!" I snarled in his ear, sinking the dagger point into his flesh. "Fly, damn you, or I'll cut your heart out!"

His wings began to thrash the air, and we rose slowly from the earth. It was a most sensational experience, but one to which I gave scant thought at the time, being so engrossed in my fury at the abduction of Altha.

When we had risen to a height of about a thousand feet, I looked for the abductors, and saw them far away, a mere group of black dots in the southern sky. After them I steered my reluctant steed.

In spite of my threats and urging for greater speed the flying dots soon vanished. Still I kept on due southward, feeling that even if I failed to overtake them, I would eventually come to the great dusky rock where legend placed their habitation.

Inspired by my poniard, my bearer made good time, considering the burden he was carrying. For hours we sped over the savannas, and by the middle of the afternoon, the landscape changed. We were flying over a forest, the first I had seen on Almuric. The trees seemed to tower to a vast height.

It was near sundown when I saw the farther limits of the forest, and in the grasslands beyond, the ruins of a city. From among these ruins smoke curled upward, and I asked my steed if his companions were cooking their evening meal there. His only answer was a snarl.

We were flying low over the forest, when a sudden uproar caused me to look down. We were just passing over a narrow glade, and in it a terrific battle was taking place. A pack of hyenas had attacked a giant unicorn-like beast, as big as a bison. Half a dozen mangled, trampled bodies attested the fury of the beast's defense, and even as I peered down, he caught the single survivor on his sword-like ivory horn, and cast it a score of feet in the air, broken and torn.

In the brief fascination of the sight, I must have involuntarily loosened my grasp on my captive. For at that instant, with a convulsive bucking heave and twist, he wrenched free and hurled me sideways. Caught off guard, I clutched vainly at empty air, and rushing earthward, crashed with a stunning impact on the loamy leaf-carpeted earth, directly in front of the maddened unicorn!

I had a dazed brief glimpse of his mountainous bulk looming over me, as his massive lowered head drove his horn at my breast. Then I lurched up on one knee, simultaneously grasping that ivory sword with my left hand and seeking to deflect it, while my right hand drove my poniard up toward the great jugular. Then there came a terrific impact against my skull, and consciousness was blotted out in darkness.

CHAPTER 6

I could have been senseless only a few minutes. When I regained consciousness my first sensation was that of a crushing weight upon my limbs and body. Struggling weakly, I found that I was lying beneath the lifeless body of a unicorn. At the instant my poniard had torn open his great jugular vein, the base of his horn must have struck my head, while the vast body collapsed upon me. Only the soft spongy ground beneath me had saved me from being crushed to a pulp. Working myself out from under that bulk was a herculean task, but eventually I accomplished it, and stood up, bruised and breathless, with the half-dried blood of the monster clotted in my hair and smearing my limbs. I was a grisly sight to look at, but I wasted no time on my appearance. My erstwhile steed was nowhere in evidence, and the circling trees limited my view of the sky.

Selecting the tallest of these trees, I climbed it as swiftly as possible, and on the topmost branches, looked out over the forest. The sun was setting. I saw that perhaps an hour's swift walk to the south, the forest thinned out and ceased. Smoke still drifted thinly up from the deserted city. And I saw my former captive just dropping down among the ruins. He must have lingered, after he had overthrown me, possibly to see if I showed any signs of life, probably to rest his wings after that long grind.

I cursed; there went my chance of stealing up on them unsuspected. Then I got a surprise. No sooner had the Yaga vanished than he reappeared, shooting up out of the city like a rocket. Without hesitation he raced off southward, speeding through the sky at a rate that left me gaping. What was the reason for his flight? If it had been his companions who were among the ruins, why had he not alighted? Perhaps he had found them gone, and was merely following them. Yet his actions seemed strange, considering the leisurely way he had approached the ruins. His flight had the earmarks of panic.

Shaking my head in puzzlement, I descended the tree and set out for the ruins as swiftly as I could make my way through the dense growth, paying no heed to the rustling in the leaves about me, and the muttering of rousing life, that grew as the shadows deepened.

Night had fallen when I emerged from the forest, but the moon was rising, casting a weird unreal glow over the plains. The ruins glimmered ghostly in the near distance. The walls were not of the rough greenish material used by the Guras. As I approached I saw they were of marble, and that fact caused a vague uneasiness to stir in my mind. I remembered legends told by the Kothans of ruined marble cities haunted by ghoulish beings. Such ruins were found in certain uninhabited places, and none knew their origin.

A brooding silence lay over the broken walls and columns as I entered the ruins. Between the gleaming white tusks and surfaces deep black shadow floated, almost liquid in its quality. From one dusky pool to the other I glided silently, sword in hand, expecting anything from an ambush by the Yagas to an attack by some lurking beast of prey. Utter silence reigned, as I had never encountered it anywhere on Almuric before. Not a distant lion roared, not a night fowl voiced its weird cry. I might have been the last survivor on a dead world.

In silence I came to a great open space, flanked by a circle of broken pillars, which must have been a plaza. Here I halted, motionless, my skin crawling.

In the midst of the plaza smoldered the dying coals of a fire over which, on spits planted in the earth, were roasting pieces of meat. The Yagas had evidently built that fire and-prepared to sup; but they had not eaten of their meal. They lay strewn about the plaza in a way to appall the hardiest.

I had never gazed on such a scene of organic devastation. Hands, feet, grinning heads, bits of flesh, entrails, clots of blood littered the whole plaza. The heads were like balls of blackness, rolled out of the shadows against the snowy marble; their teeth grinned, their eyes glimmered palely in the moonlight. *Something* had come upon the winged men as they sat about their fire and had torn them limb from limb. On the remnants of flesh were the marks of fangs, and some of the bones had been broken, apparently to get the marrow.

A cold ripple went up and down my spine. What animal but man breaks bones in that fashion? But the scattering of the bloody remnants seemed not the work of beasts; it seemed too vindictive, as if it were the work of vengeance, fury or bestial blood-thirstiness.

Where, then, was Altha? Her remains were not among those of her captors. Glancing at the flesh on the spit, the configuration of the pieces set me to shuddering. Shaken with horror, I saw that my dark suspicions were correct. It was parts of a human body the accursed Yagas had been roasting for their meal. Sick with revulsion and dread, I examined the pitiful remnants more closely, and breathed a deep sigh of relief to see the thick muscular limbs of a man, and not the slender parts of a woman. But after that I looked unmoved at the torn bloody bits that had been Yagas.

But where was the girl? Had she escaped the slaughter and hidden herself, or had she been taken by the slayers? Looking about at the towers and fallen blocks and pillars, bathed in the weird moonlight, I was aware of a distinct aura of evil, of lurking menace. I felt the glare of hidden eyes.

But I began casting about the plaza, and came upon a trail of blood drops, lying blackly in the moon, leading through a maze of drunken pillars, and for want of better occupation, I followed it. At least it might lead me to the slayers of the winged men.

I passed under the shadows of leaning pillars which dwarfed my human frame with their brute massiveness, and came into a crumbling edifice, overgrown with lichen. Through the broken roof and the gaping windows the moon poured a fungus-white light that served to make the shadows blacker. But a square of moonlight fell across the entrance of a corridor, and leading into it, I saw the sprinkle of dark clotted drops on the cracked vine-grown marble. Into the corridor I groped, and almost broke my neck on the stairs that lay within. Down them I went, and striking a level, hesitated and was about to retrace my steps when I was electrified by a sound that stopped my heart, and then sent the blood pounding madly through my veins. Through the darkness, faint and far away, sounded the call: "Esau! Esau Cairn!"

Altha! Who else could it be? Why should an icy shuddering pass over me, and the short hairs bristle at the back of my neck? I started to answer; then caution clutched my tongue. She could not know I was within hearing, surely. Perhaps she was calling as a frightened child will cry for someone far out of hearing. I went as swiftly down the black tunnel as I dared, in the direction I had heard the cry. And was gagged by a tendency toward nausea.

My groping hand encountered a doorway and I halted, sensing, as a wild thing does; a living presence of some sort near me. Straining my eyes in the pitch dark, I spoke Altha's name in a low urgent voice. Instantly two lights burned in the darkness, yellowish glows at which I stared for an instant before I realized that they were two eyes. They were broad as my hand, round and of a scintillance I cannot describe. Behind them I got a vague impression of a huge shapeless bulk. Simultaneously such a wave of instinctive fear swept over me, that I withdrew quickly into the tunnel and hastened along it in the direction I had been going. Back in the cell I heard a faint movement, like the shifting of some great pulpy mass, mingled with a soft rasping sound, as of bristles scraping against stone.

A few score paces more and I halted. The tunnel seemed endless, and besides, judging from the feel, other tunnels branched off from it in the darkness, and I had no way of knowing which was the right one. As I stood there I again heard the call: "Esau! Esau Cairn!"

Steeling myself against something, I knew not what, I set off once more in the direction of the ghostly voice. How far I went I do not know, until I stopped once more baffled. Then from nearby the voice rang out again:

"Esau! Esau CairNNNN!" It rose to a high-pitched note, trailing off into an awful burst of inhuman laughter that froze the blood in my veins.

That was not Altha's voice. I had known all the time that it was not-that it could not be. Yet the alternative was so inexplicable that I had refused to heed what my intuition affirmed and my reason denied.

Now from every direction, on every hand rose a medley of shrill demoniac voices, all shrieking my name with the mockery of devils. The tunnels that had been so silent now rang and re-echoed with strident clamor. I stood bewildered and terrified, as the damned must stand in the clamorous halls of hell. I passed through the stages of icy terror, bewildered horror, desperation, berserk fury. With a maddened roar I plunged blindly at the sounds that seemed nearest, only to collide with a solid wall, while a thousand inhuman voices rose in hideous mirth. Wheeling like a wounded bull, I charged again, this time into the mouth of another tunnel. Racing down this, mad to come to grips with my tormenters, I burst into a vast shadowy space, into which a beam of moonlight cast a ghostly shaft. And again I heard my name called, but in human tones of fear and anguish:

"Esau! Oh, Esau!"

Even as I answered the piteous cry with a savage bellow, I saw her. Altha, etched in the dim moonlight. She was stretched out on the floor, her hands and feet in the shadow. But I saw that at each outstretched member squatted a dim misshapen figure.

With a blood-thirsty yell I charged, and the darkness sprang into nauseous life, flooding my knees with tangible shapes. Sharp fangs gashed me, apish hands clawed at me. They could not halt me. Swinging my sword in great arcs that cut a path through solid masses of writhing shapes, I forged toward the girl that twisted and screamed on the floor in that square of moonlight.

I waded through a rolling, surging mire of squirming biting things that washed about me waist-high, but they could not drag me down. I reached the moonlit square, and the creatures that held Altha gave back before the whistling menace of my sword edge, and the girl sprang up and clung to me. Even as the shadowy horde rolled in to drag us down I saw a crumbling stair leading up, and I thrust her upon it, wheeling to cover her retreat.

It was dark on the stairs, though they led up into a chamber flooded with light through a broken roof. That battle was fought in utter darkness, with only my senses of feeling and hearing to guide my strokes. And it was fought in silence, too, except for my panting, and the whir and crunch of my blade.

Up that drunken stair I backed, battling every inch of the way, the skin between my shoulders crawling with the expectancy of an attack from the rear. If they had come upon us from above, we had been lost, but evidently all were below me. What manner of creatures I was fighting I did not know, except that they were taloned and fanged. Otherwise, from the feel of them, they were stunted and misshapen, furry and apish.

When I came out into the chamber above the tunnels I could see little more. The moonlight streaming through the broken roof made only a white shaft in the darkness. I could only make out vague forms in the dimness about me-a heaving, writhing and lashing of shadows, that surged up against me, clawing and tearing, and fell back beneath my lashing sword.

Thrusting Altha behind me, I backed across that shadowy chamber toward a wide rift that showed in the crumbling wall, reeling and stumbling in the whirlpool of battle that swirled and eddied about me. As I reached the rift through which Altha had already slipped, there was a concerted rush to drag me down. Panic swept over me at the thought of being pulled down in that shadowy room by that dim horde. A blasting burst of fury, a gasping, straining plunge, and I catapulted through the rift, carrying half a dozen attackers with me.

Reeling up, I shook the clinging horrors from my shoulders as a bear might shake off wolves, and bracing my feet slashed right and left. Now for the first time I saw the nature of my foes.

The bodies were like those of deformed apes, covered with sparse dirty white fur. Their heads were doglike, with small close-set ears. But their eyes were those of serpents-the same venomous steady lidless stare.

Of all the forms of life I had encountered on that strange planet, none filled me with as much loathing as these dwarfish monstrosities. I backed away from the mangled heap on the earth, as a nauseous flood poured through the rift in the wall.

The effect of those vermin emerging from that broken wall was almost intolerably sickening; the suggestion was that of maggots squirming out of a cracked and bleached skull.

Turning, I caught Altha up in one arm and raced across the open space. They followed fleetingly, running now on all fours, and now upright like a man. And suddenly they broke out into their hellish laughter again, and I saw we were trapped. Ahead of me were more emerging from some other subterranean entrance. We were cut off.

A giant pedestal, from which the column had been broken, stood before us. With a bound I reached it, set the girl on the jagged pinnacle, and wheeled on the lower base to take such toll of our pursuers as I might. Blood streaming from a score of gashes trickled down the pedestal on which I stood, and I shook my head violently to rid my eyes of blinding sweat.

They ringed me in a wide semicircle, deliberate now that their prey seemed certain, and I cannot recall a time when I was more revolted by horror and disgust, than when I stood with my back to that marble pillar and faced those verminous monsters of the lower world.

Then my attention was caught by a movement in the shadows under the wall through which we had just come. Something was emerging from the rift-something huge and black and bulky. I caught the glitter of a yellowish spark. Fascinated, I watched, even while the furred devils were closing in. Now the thing had emerged entirely from the rift. I saw it crouching in the shadow of the wall, a squat mass of blackness from which glimmered a pair of yellowish lights. With a start I recognized the eyes I had seen in the subterranean cell.

With a clamor of fiendish yells the furry devils rushed in, and at the same instant the unknown creature ran out into the moonlight with surprising speed and agility. I saw it plainly then-a gigantic spider, bigger than an ox. Moving with the swiftness characteristic of its breed, it was among the dog-heads before the first had felt my lifted sword. An awful scream rose from its first victim, and the rest, turning, broke and fled shrieking in all directions. The monster raged among them with appalling quickness and ferocity. Its huge jaws crunched their skulls, its dripping mandibles skewered them, it crushed their bodies by its sheer weight. In an instant the place was a shambles, inhabited only by the dead and dying. Crouching among its victims, the great black hairy thing fixed its horribly intelligent eyes on me.

I was the one it was trailing. I had awakened it underground, and it had followed the scent of the dried blood on my sandals. It had slaughtered the others simply because they stood in its way.

As it crouched on its eight bent legs, I saw that it differed from Earthly spiders not only in size, but in the number of its eyes and the shape of its jaws. Now Altha screamed as it ran swiftly toward me.

But where the fangs and claws of a thousand beast-things were futile against the venom dripping from those black mandibles, the brain and thews of a single man prevailed. Catching up a heavy block of masonry, I poised it for an instant, and then hurled it straight into the onrushing bulk. Full among those branching hairy legs it crushed, and a jet of nauseous green stuff gushed into the air from the torn torso. The monster, halted in his rush, writhed under the pinning stone, cast it aside and staggered toward me again, dragging broken legs, its eyes glittering hellishly. I tore another missile from the crumbling stone, and another and another, raining huge chunks of marble on the writhing horror until it lay still in a ghastly mess of squirming hairy black legs, entrails and blood.

Then catching Altha in my arms, I raced away through the shadows of monolith and tower and pillar, nor did I halt until the city of silence and mystery lay behind us, and we saw the moon setting across the broad waving grasslands.

No word had passed between us since I had first come upon the girl in that ghoulish tunnel. Now when I looked down to speak to her I saw her dark head drooping against my arm; her white face was upturned, her eyes shut. A quick throb of fear went through me, but a swift examination showed me that she had merely fainted. That fact showed the horror of what she had been through. The women of Koth do not faint easily.

I laid her at full length on the turf, and gazed at her helplessly, noting, as if for the first time, the white firmness of her slender limbs, the exquisite molding of her supple figure. Her dark hair fell in thick glossy clusters about her alabaster shoulders, a strap of her tunic, slipped down, revealed her firm, pink-tipped young breasts. I was aware of a vague unrest that was almost a pain.

Altha opened her eyes and looked up at me. Then her dark eyes flared with terror, and she cried out and clutched at me frantically. My arms closed about her instinctively, and within their iron-thewed clasp I felt the pulsating of her lithe body, the wild fluttering of her heart.

"Don't be afraid." My voice sounded strange, scarcely articulate. "Nothing is going to harm you."

I could feel her heart resuming its normal beat, so closely she clung to me, before her quick pants of fright ceased. But for a while she lay in my arms, looking up at me without speaking, until, embarrassed, I released her and lifted her to a sitting position on the grass.

"As soon as you feel fit," I said, "we'll put more distance between us and-that." I jerked my head in the direction of the distant ruins.

"You are hurt," she exclaimed suddenly, tears filling her eyes. "You are bleeding! Oh, I am to blame. If I had not run away~" She was weeping now in earnest, like any Earthly girl.

"Don't worry about these scratches," I answered, though privately I was wondering if the fangs of the vermin were venomous. "They are only flesh wounds. Stop crying, will you?"

She obediently stifled her sobs, and naively dried her eyes with her skirt. I did not wish to remind her of her horrible experience, but I was curious on one point.

"Why did the Yagas halt at the ruins?" I asked. "Surely they knew of the things that haunt such cities."

"They were hungry," she answered with a shudder. "They had captured a youth-they dismembered him alive, but never a cry for mercy they got, only curses. Then they roasted-" She gagged, smitten with nausea.

"So the Yagas are cannibals." I muttered.

"No. They are devils. While they sat about the fire the dog-heads fell upon them. I did not see them until they were on us. They swarmed over the Yagas like jackals over deer. Then they dragged me into the darkness. What they meant to do, Thak only knows. I have heard-but it is too obscene to repeat."

"But why did they shriek my name?" I marveled.

"I cried it aloud in my terror," she answered. "They heard and mimicked me. When you came, they knew you. Do not ask me how. They too are devils."

"This planet is infested with devils," I muttered. "But why did you call on me, in your fright, instead of your father?"

She colored slightly, and instead of answering, began pulling her tunic straps in place.

Seeing that one of her sandals had slipped off, I replaced it on her small foot, and while I was so occupied she asked unexpectedly: "Why do they call you Ironhand? Your fingers are hard, but their touch is as gentle as a woman's. I never had men's fingers touch me so lightly before. More often they have hurt me."

61

I clenched my fist and regarded it moodily-a knotted iron mallet of a fist. She touched it timidly.

"It's the feeling behind the hand." I answered. "No man I ever fought complained that my fists were gentle. But it is my enemies I wish to hurt, not you."

Her eyes lighted. "You would not hurt me? Why?"

The absurdity of the question left me speechless.

CHAPTER 7

It was past sunrise when we started back on the long trek toward Koth, swinging far to the west to avoid the devil city from which we had escaped. The sun came up unusually hot. The air was breathless, the light morning wind blew fitfully, and then died down entirely. The always cloudless sky had a faint copperish tint. Altha eyed that sky apprehensively, and in answer to my question said she feared a storm. I had supposed the weather to be always clear and calm and hot on the plains, clear and windy and cold in the hills. Storms had not entered into my calculation.

The beasts we saw shared her uneasiness. We skirted the edge of the forest, for Altha refused to traverse it until the storm had passed. Like most plains-dwellers, she had an instinctive distrust of thick woods. As we strode over the grassy undulations, we saw the herds of grazers milling confusedly. A drove of jumping pigs passed us, covering the ground with gargantuan bounds of thirty and forty feet. A lion started up in front of us with a roar, but dropped his massive head and slunk hurriedly away through the tall grass.

I kept looking for clouds, but saw none. Only the copperish tint about the horizons grew, discoloring the whole sky. It turned from light color to dull bronze, and from bronze to black. The sun smoldered for a little like a veiled torch, veining the dusky dome with fire, then it was blotted out. A tangible darkness seemed to hover an instant in the sky, then rush down, cloaking the world in utter blackness, through which shone neither sun, moon, nor stars. I had never guessed how impenetrable darkness could be. I might have been a blind, disembodied spirit wandering through unlighted space, but for the swish of the grasses under my feet, and the soft warm contact of Altha's body against mine. I began to fear we might fall into a river, or blunder against some equally blind beast of prey.

I had been making for a mass of broken boulders, such a formation as occasionally occurs on the plains. Darkness fell before we reached them, but groping on, I stumbled upon a sizable rock, and placing my back to it, drew Altha against it and stood sheltering her with my body as well as I could. Out on the dark plains breathless silence alternated with the sounds of varied and widespread movement-rustling of grass, shuffle of padded hoofs, weird lowing and low-pitched roaring. Once a vast herd of some sort swept by us, and I was thankful for the protection of the boulders that kept us from being trampled. Again all sounds ceased and the silence was as complete as the darkness. Then from somewhere came a weird howling.

"What's that?" I asked uneasily, unable to classify it.

"The wind!" she shivered, snuggling closer to me.

It did not blow with a steady blast; here and there it swept in mad fitful gusts. Like lost souls it wailed and moaned. It ripped the grasses near us, and finally a puff of it struck us squarely, knocking us off our feet and bruising us against the boulder behind us. Just that one abrupt blast, like a buffet from an unseen giant's fist.

As we regained our feet I froze. Something was passing near our refuge-something mountain-huge, beneath whose tread the earth trembled. Altha caught me in a desperate clutch, and I felt the pounding of her heart. My hair prickled with nameless fear. The *thing* was even with us. It halted, as if sensing our presence. There was a curious leathery sound, as of the movement of great limbs. Something waved in the air above us; then I felt a touch on my elbow. The same object touched Altha's bare arm, and she screamed, her taut nerves snapping.

Instantly our ears were deafened by an awful bellow above us, and something swept down through the darkness with a clashing of gigantic teeth. Blindly I lashed out and upward, feeling my sword-edge meet tangible substance. A warm liquid spurted along my arm, and with another terrible roar, this time more of pain than rage, the invisible monster shambled away, shaking the earth with its tread, dimming the shrieking wind with its bellowing.

"What was it, in God's name?" I panted.

"It was one of the Blind Ones," she whispered. "No man has ever seen them; they dwell in the darkness of the storm. Whence they come, whither they go, none knows. But look, the darkness melts."

"Melts" was the right word. It seemed to shred out, to tear in long streamers. The sun came out, the sky showed blue from horizon to horizon. But the earth was barred fantastically with long strips of darkness, tangible shadows floating on the plain, with broad spaces of sunlight between. The scene might have been a dream landscape of an opium-eater. A hurrying deer flitted across a sunlight band and vanished abruptly in a broad streamer of black; as suddenly it flashed into light and sight again. There was no gradual shaking into darkness; the borders of the torn strips of blackness were as clear-cut and definite as ribbons of ebony on a background of gold and emerald. As far as I could see, the world was stripped and barred with those black ribbons. Sight could not pierce them, but they were thinning, dividing, vanishing.

Directly before one of the streamers of darkness ripped apart and disappeared, revealing the figure of a man-a hairy giant, who stood glaring at me, sword in hand, as surprised as I. Then several things happened all at once. Altha screamed: "A Thugran!" the stranger leaped and slashed, and his sword clanged on my lifted blade.

I have only a brief chaotic memory of the next few seconds. There was a whirl of strokes and parries, a brief clanging of steel; then my sword-point sank under his heart and stood out behind his back. I wrenched the blade free as he sank down, and stood glaring down at him bewilderedly. I had secretly wondered what the outcome would be when I was called upon to face a seasoned warrior with naked steel. Now it had occurred and was over with, and I was absolutely unable to remember how I had won. It had been too fast and furious for conscious thought; my fighting instincts had acted for me.

A clamor of angry cries burst on me, and wheeling I saw a score of hairy warriors swarming out from among the rocks. It was too late to flee. In an instant they were on me, and I was the center of a whirling, flashing, maelstrom of swords. How I parried them even for a few seconds I cannot say. But I did, and even had the satisfaction of feeling my blade grate around another, and sever the wielder's shoulder bone. A moment later one stooped beneath my thrust and drove the spear through the calf of my leg. Maddened by the pain, I dealt him a stroke that split his skull to the chin, and then a carbine stock descended on my head. I partially parried the blow, else it had smashed my skull. But even so, it beat down on my crown with thunderous and murderous impact, and the lights went out.

I came to with the impression that I was lying in a small boat which was rocking and tossing in a storm. Then I discovered that I was bound hand and foot, and being borne on a litter made of spear-shafts. Two huge warriors were bearing me between them, and they made no effort to make the traveling any easier for me. I could see only the sky, the hairy back of the warrior in front of me, and by drawing back my head the bearded face of the warrior behind. This person, seeing my eyes open, growled a word to his mate, and they promptly dropped the litter. The jolt set my damaged head to throbbing, and the wound in my leg to hurting abominably.

"Logar!" bawled one of them. "The dog is conscious. Make him walk, if you must bring him to Thugra. I've carried him far as I'm going to."

I heard footsteps, and then above me towered a giant form and a face that seemed familiar. It was a fierce, brutal face, and from the corner of the snarling mouth to the rim of the square jaw, ran a livid scar.

"Well, Esau Cairn," said this individual, "we meet again."

I made no reply to this obvious comment.

"What?" he sneered, "do you not remember Logar the Bonecrusher, you hairless dog?"

He punctuated his remarks by a savage kick in my ribs. Somewhere there rang out a feminine shriek of protest, the sound of a scuffle, and Altha broke through the ring of warriors and fell to her knees beside me.

"Beast!" she cried, her beautiful eyes blazing. "You kick him when he is helpless, when you would not dare face him in fair battle."

"Who let this Kothan cat loose?" roared Logar. "Thal, I told you to keep her away from this dog."

"She bit my hand," snarled the big warrior, striding forward, and shaking a drop of blood from his hairy paw. "I'd as soon try to hold a spitting wildcat."

"Well, haul him to his feet." directed Logar. "He walks the rest of the way."

"But he is wounded in the leg!" wailed Altha. "He cannot walk."

"Why don't you finish him here?" demanded one of the warriors.

"Because that would be too easy!" roared Logar, red lights flickering in his blood-shot eyes. "The thief struck me foully with a stone, from behind, and stole my poniard."-here I saw that he was wearing it once more at his girdle. "He shall go to Thugra, and there I'll take my time about killing him. Drag him up!"

They loosened my legs, none too gently, but the wounded one was so stiff I could hardly stand, much less walk. They encouraged me with blows, kicks, and prods from spears and swords, while Altha wept in helpless fury, and at last turned on Logar.

"You are both a liar and a coward!" she screamed. "He did not strike you with a stone-he beat you down with his naked fists, as all men know, though your slaves dare not acknowledge it-"

Logar's knotty fist crashed against her jaw, knocking her off her feet, to fall in a crumpled heap a dozen feet away. She lay without moving, blood trickling from her lips. Logar grunted in savage satisfaction, but his warriors were silent. Moderate corporal correction for women was not unknown among the Guras, but such excessive and wanton brutality was repugnant to any warrior of average decency. So Logar's braves looked glum, though they made no verbal protest.

As for me, I went momentarily blind with the red madness of fury that swept over me. With a blood-thirsty snarl I jerked convulsively, upsetting the two men who held me; so we all went down in a heap. The other Thugrans came and boosted us up, glad to vent their outraged feelings on my carcass, which they did lustily, with sandal heels and sword hilts. But I did not feel the blows that rained upon me. The whole world was swimming red to my sight, and speech had utterly failed me. I could only snarl bestially as I tore in vain at the thongs which bound me. When I lay exhausted, my captors hauled me up and began beating me to make me walk.

"You can beat me to death," I snarled, finding my voice at last, "but I won't move until some of you see to the girl."

"The slut's dead," growled Logar.

"You lie, you dog!" I spat. "You miserable weakling, you couldn't hit hard enough to kill a new-born babe!"

Logar bellowed in wordless fury, but one of the others, panting from his exertions of hammering me, stepped over to Altha, who was showing signs of life.

"Let her lie!" roared Logar.

"Go to the devil!" snarled his warrior. "I love her no more than you do, but if bringing her along will make that smooth-skinned devil walk of his own accord, I'll bring her, if I have to carry her all the way. He's not human; I've pummeled him till I'm ready to drop dead, and he's in better shape than I."

So Altha, wobbly on her legs and very groggy, accompanied us as we marched to Thugra.

We were on the road several days, during which time walking was agony to my wounded leg. Altha persuaded the warriors to let her bandage my wounds, and but for that I very probably should have died. I was marked in many places by the gashes received in the haunted ruins, battered and bruised from head to foot by the beating the Thugrans gave me. Just enough food and water was given me to keep me alive. And so, dazed, weary, harassed by thirst and hunger, crippled, stumbling along over those endless rolling plains, I was even glad at last to see the walls of Thugra looming in the distance, even though I knew they spelled my doom. Altha had not been badly treated on the march, but she had been prevented from giving me aid and comfort, beyond bandaging my wounds, and all through the nights, waking from the beastlike sleep of utter exhaustion, I heard her sobbing. Among the hazy, tortured impression of that dreary trek, that stands out most clearly-Altha sobbing in the night, terrible with loneliness and despair in the immensity of shadowed world and moaning darkness.

And so we came to Thugra. The city was almost exactly like Koth-the same huge tower-flanked gates, massive walls built of rugged green stone, and all. The people, too, differed none in the main essentials from the Kothans. But I found that their government was more like an absolute monarchy than was Koth's. Logar was a primitive despot, and his will was the last power. He was cruel, merciless, lustful and arrogant. I will say this for him: he upheld his rule by personal strength and courage. Thrice during my captivity in Thugra I saw him kill a rebellious warrior in hand-to-hand combat-once with his naked hands against the other's sword. Despite his faults, there was force in the man, a gusty, driving, dynamic power that beat down opposition with sheer brutality. He was like a roaring wind, bending or breaking all that stood before him.

Possessed of incredible vitality, he was intensely vain of his physical prowess-in which, I believe, his superiority of personality was rooted. That was why he hated me so terribly. That was why he lied to his people and told them that I struck him with a stone. That was why, too, he refused to put the matter to test. In his heart lurked fear-not of any bodily harm I might do him, but fear lest I overcome him again, and discredit him in the eyes of his subjects. It was his vanity that made a beast of Logar.

I was confined in a cell, chained to the wall. Logar came every day to curse and taunt me. It was evident that he wished to exhaust all mental forms of torment before he proceeded to physical torture. I did not know what had become of Altha. I had not seen her since first we entered the city. He swore that he had taken her to his palace and described to me with great detail the salacious indignities to which he swore he subjected her. I did not believe him, for I felt he would be more likely to bring her to my cell and torture her before me. But the fury into which his obscene narrations threw me could not have been much more violent if the scenes he described had been enacted before me.

It was easy to see that the Thugrans did not relish Logar's humor, for they were no worse than other Guras, and all Guras possess, as a race, an innate decency in regard to women. But Logar's power was too complete for any to venture a protest. At last, however, the warrior who brought me food told me that Altha had disappeared immediately after we reached the city, and that Logar was searching for her, but unable to find her. Apparently she had either escaped from Thugra, or was hiding somewhere in the city.

And so the slow days crawled by.

CHAPTER 8

It was midnight when I awoke suddenly. The torch in my cell was flickering and guttering. The guard was gone from my door. Outside, the night was full of noise. Curses, yells, and shots mingled with the clash of steel, and over all rose the screaming of women. This was accompanied by a curious thrashing sound in the air above. I tore at my bonds, mad to know what was happening. There was fighting in the city, beyond the shadow of a doubt, but whether civil war or alien invasion, I could not know.

Then quick light steps sounded outside, and Altha ran swiftly into my cell. Her hair was in wild disorder, her scanty garment torn, her eyes ablaze with terror.

"Esau!" she cried. "Doom from the sky has fallen on Thugra! The Yagas have descended on the city by the thousands! There is fighting in the streets and on the house tops-the gutters are running red, and the streets are strewn with corpses! Look! The city is burning!"

Through the high-set barred windows I saw a smoldering glow. Somewhere sounded the dry crackling of flames. Altha was sobbing as she fumbled vainly at my bonds. That day Logar had begun the physical torture, and had had me hauled upright and suspended from the roof by a rawhide thong bound about my wrists, my toes just touching a huge block of granite. But Logar had not been so wise. They had used a new thong of hide, and it had stretched, allowing my feet to rest on the block, in which position I had suffered no unbearable anguish, and had even fallen asleep, though naturally the attitude was not conducive to great comfort.

As Altha worked futilely to free me, I asked her where she had been, and she answered that she had slipped away from Logar when we had reached the city, and that kind women, pitying her, had hidden and fed her. She had been waiting for an opportunity to aid me in escaping. "And now," she wailed, wringing her hands, "I can do nothing! I cannot untie this wretched noose!"

"Go find a knife!" I directed. "Quick!"

Even as she turned, she cried out and shrank back, trembling, as a terrible figure lurched through the door.

It was Logar, his mane and beard matted and singed, the hair on his great

70

breast crisped and blackened, blood streaming from his limbs. His blood-shot eyes glared madness as he reeled toward me, lifting the poniard I had taken from him so long before.

"Dog!" he croaked. "Thugra is doomed! The winged devils drop from the skies like vultures on a dead ox! I have slain until I die of weariness, yet still they come. But I remembered you. I could not rest easy in Hell, knowing you still lived. Before I go forth again to die, I'll send you before me!"

Altha shrieked and ran to shield me, but he was before her. Rising on his toes he caught at my girdle, lifting the poniard on high. And as he did so, I drove my knee with terrific force up against his jaw. The impact must have broken his bull-neck like a twig. His shaggy head shot back between his shoulders, his bearded chin pointing straight up. He went down like a slaughtered ox, his head crashing hard on the stone floor.

A low laugh sounded from the doorway. Etched in the opening stood a tall ebony shape, wings half lifted, a dripping scimitar in a crimsoned hand. Limned in the murky red glare behind him, the effect was that of a black-winged demon standing in the flame-lit door of Hell. The passionless eyes regarded me enigmatically, flitted across the crumpled form on the floor, then rested on Altha, cowering at my feet.

Calling something over his shoulder, the Yaga advanced into the room, followed by a score of his kind. Many of them bore wounds, and their swords were notched and dripping.

"Take them," the first comer indicated Altha and myself.

"Why the man?" demurred one.

"Who ever saw a white man with blue eyes before? He will interest Yasmeena. But be careful. He has the thews of a lion."

One of them grasped Altha's arm and dragged her away, struggling vainly and twisting her head to stare back at me with terrified eyes, and the others from a safe distance cast a silken net about my feet. While my limbs were so enmeshed, they seized me, bound me with silken cords that a lion could not have broken, and cut the thong by which I was suspended. Then two of them lifted me and bore me out of the cell. We emerged into a scene of frenzy in the streets.

The stone walls were of course immune to flame, but the woodwork of the buildings was ablaze. Smoke rolled up in great billowing clouds, shot and veined by tongues of flame, and against this murky background black shapes twisted and contorted like figments of nightmare. Through the black clouds shot what appeared to be blazing meteors, until I saw they were winged men bearing torches.

In the streets, among falling sparks and crashing walls, in the burning buildings, on the roofs, desperate scenes were being hideously enacted. The men of Thugra were fighting with the fury of dying panthers. Any one of them was more than a match for a single Yaga, but the winged devils far outnumbered them, and their fiendish agility in the air balanced the superior strength and courage of the ape-men. Swooping down through the air, they slashed with their curved swords, soaring out of reach again before the victim could return the stroke. When three or four devils were striking thus at a single enemy, the butchery was certain and swift. The smoke did not seem to bother them as it did their human adversaries. Some, perched on points of vantage, bent bows and sent arrows singing down into the struggling masses in the streets.

The killing was not all on one side. Winged bodies as well as hairy shapes lay strewn in the blood-splashed streets. Carbines cracked and more than a few flying fiends crashed earthward in a frantic thrashing of wings. Madly lashing swords found their target, and when the desperate hands of a Gura closed on a Yaga, that Yaga died horribly.

But by far the greater slaughter was among the Thugrans. Blinded and half strangled, most of their bullets and arrows went wild. Outnumbered and bewildered by the hawk-like tactics of their merciless foes, they fought vainly, were cut down or feathered with arrows.

The main object of the Yagas seemed to be women captives. Again and again I saw a winged man soar up through the whirling smoke, gripping a shrieking girl in his arms.

Oh, it was a sickening sight! I do not believe that the utter barbarism and demoniac cruelty of the scene could be duplicated on Earth, vicious as its inhabitants can be at times. It was not like humans fighting humans, but like members of two different forms of life at war, utterly without sympathy or any common plane of understanding.

But the massacre was not complete. The Yagas were quitting the city they had ruined, sweeping up into the skies laden with naked writhing captives. The survivors still held the streets, and fired blindly at the departing victors, evidently preferring to risk killing their captives rather than to let them be carried to the fate that awaited them.

I saw a knot of perhaps a hundred struggling fighters slashing and gasping on the highest roof in the city, the Yagas to tear away and escape, the Guras to drag them down. Smoke billowed about them, flames caught at their hair; then with a thunderous roar the roof fell in, bearing victors and vanquished alike to a fiery death. The deafening thunder of the devouring flames was in my ears as my captors whirled me through the air away from the reeking city of Thugra.

When my dazed faculties adjusted themselves sufficiently for me to take note of my surroundings, I found myself sailing through the sky at terrific speed, while below, above and about me sounded the steady beat of mighty wings. Two Yagas were bearing me with perfect ease, and I was in the midst of the band, which was flying southward in a wedge-shaped formation, like that of wild geese. There were fully ten thousand of them. They darkened the morning sky, and their gigantic shadow swept over the plain beneath them as the sun rose.

We were flying at an altitude of about a thousand feet. Many of the winged men bore girls and young women, and carried them with an ease that spoke of incredible wing-power. No match in sheer muscularity for the Guras, yet these winged devils have unbelievable powers of endurance in the air. They can fly for hours at top speed, and in the wedge formation, with unburdened leaders cleaving the air ahead of them, can carry weights almost equal their own at almost the same velocity.

We did not pause to rest or eat until nightfall, when our captors descended to the plain, where they built fires and spent the night. That night lives in my memory as one of the greatest horrors I have ever endured. We captives were given no food, but the Yagas ate. And their food was their miserable captives. Lying helpless, I shut my eyes to that butchery, wished that I were deaf that I might not hear the heart-rending cries. The butchery of men I can endure, in battle, even in red massacre. The wanton slaughter of helpless women who can only shriek for mercy until the knife silences their wails, that is more than I can stand. Nor did I know but that Altha was among those chosen for the grisly feast. With each hiss and crunch of the beheading blade I winced, seeing in fancy her lovely dark head roll on the blood-soaked ground. For what was going on at the other fires I could not know.

After it was over and the gorged demons lay about the fires in slumber, I lay sick at heart, listening to the roaring of the prowling lions, and reflecting how kinder and gentler is any beast, than any thing molded in the form of man. And out of my sick horror grew a hate that steeled me for whatever might come, in the grim determination to ultimately repay these winged monsters for all the suffering they had inflicted.

Dawn was only a hint in the sky when we took the air again. There was no morning meal. I was to learn that the Yagas ate only at intervals, gorging

73

themselves to capacity every few days. After several hours hurtling over the usual grasslands, we came suddenly in sight of a broad river spanning the savannas from horizon to horizon, fringed on the northern bank by a narrow belt of forest. The waters were of a curious purple, glimmering like watered silk. On the farther bank appeared a tall thin tower of a black shiny material that glittered like polished steel.

As we whirled over the river I saw that it was rushing with terrific velocity. Its roar came up to us, and I saw the seething of eddying whirlpools in its racing current. Crossing the stream at the point where the tower stood, reared numbers of huge stones, among which the waters foamed and thundered. Looking down at the tower, I saw half a dozen winged men on the battlemented roof, who tossed up their arms as if hailing our captors. From the river southward stretched desert- bare, dusty, grayish, strewn occasionally with bleached bones here and there. Far away on the horizon I saw a giant black bulk growing in the sky.

It stood out boldly as we raced toward it. In a few hours we had reached it, and I was able to make out all its details. It was a gigantic block of black basaltlike rock rising sheer out of the desert, a broad river flowing about its feet, its summit crowned with black towers, minarets and castles. It was no myth, then, but a fantastic reality-Yugga, the Black City, the stronghold of the winged people.

The river, cutting through the naked desert, split on that great rock and passed about it on either side, forming a natural moat. On every side but one the waters lapped the sheer walls of the cliffs. But on one side a broad beach had been formed, and there stood another town. Its style of architecture was very different from that of the edifices on the rock. The houses were mere stone huts, squat, flat-roofed, and one-storied. Only one building had any pretensions-a black temple-like edifice built against the cliff wall. This lower town was protected by a strong stone wall built about it at the water's edge, and connecting at each end with the cliff behind the town.

I saw the inhabitants, and saw that they were neither Yagas nor Guras. They were short and squat of build, and of a peculiar blue color. Their faces, while more like those of Earthly humans than were those of the Gura males, lacked the intelligence of the latter. The countenances were dull, stupid and vicious, the women being little more prepossessing than the men. I saw these curious people, not only in their town at the foot of the cliff, but at work in fields along the river.

I had little opportunity for observing them, however, since the Yagas swept straight up to the citadel, which towered five hundred feet above the river. I was bewildered by the array of battlements, pinnacles, minarets and roof gardens that met my gaze, but got the impression that the city on the rock was built like one huge palace, each part connected with the rest. Figures lounging on couches on the flat roofs lifted themselves on elbow, and from scores of casements the faces of women looked at us as we sank down on a broad flat roof that was something like a landing-field. There many of the winged men dispersed, leaving the captives guarded by three or four hundred warriors, who herded them through a gigantic door. There were about five hundred of these wretched girls, Altha among them. I was carried, still bound, along with them. By this time my whole body was numb from having circulation cut off so long, but my mind was intensely active.

We traversed a stairway down which half a dozen elephants could have stalked abreast, and came into a corridor of corresponding vastness. Walls, stair, ceiling and floor were all of the gleaming black stone, which I decided had been cut out of the rock on which Yugga was built, and highly polished. So far I had seen no carvings, tapestries, or any attempt at ornamentation; yet it could not be denied that the effect of those lofty walls and vaulted ceilings of polished ebony was distinctly one of splendor. There was an awe-inspiring majesty about the architecture which seemed incongruous, considering the beastliness of the builders. Yet the tall black figures did not seem out of place, moving somberly through those great ebony halls. The Black City-not alone because its walls were dusky hued did humans give it that sinister name.

As we passed through those lofty halls I saw many of the inhabitants of Yugga. Besides the winged men, I saw, for the first time, the women of the Yagas. Theirs was the same lithe build, the same glossy black skin, the same faintly hawklike cast of countenance. But the women were not winged. They were clad in short silken skirts held up with jewel-crusted girdles, and in filmy sashes bound about their breasts. But for the almost intangible cruelty of their faces, they were beautiful. Their dusky features were straight and clear-cut, their hair was not kinky.

I saw other women, hundreds of the black-haired, white-skinned daughters of the Guras. But there were others: small, dainty, yellow-skinned girls, and copper-colored women-all, apparently, slaves to the black people. These women were something new and unexpected. All the fantastic forms of life I had encountered so far had been mentioned in tales or legends of the Kothans. The dog-heads, the giant spider, the winged people with their black citadel and their blue-skinned slaves-all these had been named in legendry, at least. But no man or woman of Koth had ever spoken of women with yellow or copper skins. Were these exotic prisoners from another planet, just as I was from an alien world?

While meditating the matter I was carried through a great bronze portal at which stood a score of winged warriors on guard, and found myself with the captive girls in a vast chamber, octagonal in shape, the walls hung with dusky tapestries. It was carpeted with some sort of rich fur-like stuff, and the air was heavy with perfumes and incense.

Toward the back of the chamber, broad steps of beaten gold led up to a fur-covered dais, on which lounged a young black woman. She alone, of all the Yaga women, was winged. She was dressed like the rest, wearing no ornaments except her gem-crusted girdle, from which jutted a jeweled dagger hilt. Her beauty was marvelous and disquieting, like the beauty of a soulless statue. I sensed that of all the inhuman denizens of Yugga, she was least human. Her brooding eyes spoke of dreams beyond the boundaries of human consciousness. Her face was the face of a goddess, knowing neither fear nor mercy.

Ranged about her couch in attitudes of humility and servitude were twenty naked girls, white-, yellow- and copper-skinned.

The leader of our captors advanced toward the royal dais, and bowing low, at the same time extending his hands, palms down and fingers spread wide, he said: "Oh, Yasmeena, Queen of the Night, we bring you the fruits of conquest."

She raised herself on her elbow, and as her terribly personal gaze passed over her cringing captives, a shudder swept across their ranks as a wind passes over rows of wheat. From earliest childhood Gura girls were taught, by tales and tradition, that the worst fate that could befall them was to be captured by the people of the Black City. Yugga was a misty land of horror, ruled by the archfiend Yasmeena. Now those trembling girls were face to face with the vampire herself. What wonder that many of them fainted outright?

But her eyes passed over them and rested on me, where I stood propped up between a couple of warriors. I saw interest grow in those dark luminous eyes, and she spoke to the chief:

"Who is that barbarian, whose skin is white, yet almost as hairless as ours, who is clad like a Gura, and yet unlike them?"

"We found him a captive among the Thugrans, oh mistress of Night," he answered. "Your majesty shall herself question him. And now, oh dark beauty, be pleased to designate the miserable wenches who shall serve your loveliness, that the rest may be apportioned among the warriors who made the raid."

Yasmeena nodded, her eyes still on me, and with a few waves of her hand she indicated a dozen or so of the handsomest girls, among these being Altha. They were drawn aside, and the rest were herded out.

Yasmeena eyed me a space without speaking, and then said to him who appeared to be her major-domo: "Gotrah, this man is weary and stained with travel and captivity, and there is an unhealed wound in his leg. The sight of him, as he now is, offends me. Take him away, let him bathe and eat and drink, and let his leg be bandaged. Then bring him to me again."

So my captors with a weary sigh, heaved me up again, and carried me from the royal chamber, down a winding corridor, along a flight of stairs, and halted finally in a chamber where a fountain bubbled in the floor. There they fastened gold chains to my wrists and ankles and then cut the cords that bound me. In the excruciating pain of the returning circulation, I scarcely noticed when they splashed me in the fountain, bathing the sweat, dirt and dried blood from my limbs and body, and clad me in a new loincloth of scarlet silk. They likewise dressed the wound in my calf, and then a copper-skinned slave-girl entered with gold vessels of food. I would not touch the meat, what with my grisly suspicions, but I ate ravenously of the fruits and nuts, and drank deeply of a green wine, which I found most delicious and refreshing.

After that I felt so drowsy that I sank down on a velvet couch and passed instantly into deep slumber, from which I was roused by someone shaking me. It was Gotrah bending over me with a short knife in his hand; and, all my wild instincts aroused, I did my best to brain him with my clenched fist, and failed only because of the chain on my wrist. He recoiled, cursing.

"I have not come to cut your throat, barbarian," he snapped, "though nothing would please me better. The Kothan girl has told Yasmeena that it is your habit to scrape the hair from your face, and it is the Queen's desire to see you thus. Here, take this knife and scrape yourself. It has no point, and I will be careful to stay out of your reach. Here is a mirror."

Still half asleep-by which I believe the green wine was drugged, though for what reason I cannot say-I propped the silver mirror up against the wall, and went to work on my beard, which had reached no mean proportions during my captivities. It was a dry shave, but my skin is as durable as tanned leather, and the knife had an edge keener than I ever found on an Earthly razor. When I had finished, Gotrah grunted at my changed appearance and demanded the knife again. As there was no point in retaining it, it being useless as a weapon, I threw it at him, and immediately fell asleep again.

The next time I awoke naturally, and rising, took in my surroundings more minutely. The chamber was unadorned, furnished only with the couch, a small ebony table, and a fur-covered bench. There was a single door, which was closed and doubtless bolted on the outside, and one window. My chains were fastened to a gold ring in the wall behind the couch, but the strand that linked me to it was long enough to allow me to take a few steps to the fountain, and to the window. This window was barred with gold, and I looked out over flat roofs, at towers and minarets which limited my view.

So far the Yagas had treated me well enough; I wondered how Altha was faring, and if the position of member of the Queen's retinue carried any special privileges or safety.

Then Gotrah entered again, with half a dozen warriors, and they unlocked my chain from the wall and escorted me down the corridor, up the winding stair. I was not taken back to the great throne chamber, but to a smaller room high up in a tower. This room was so littered with furs and cushions that it was almost stuffed. I was reminded of the soft, padded nest of a spider, and the black spider was there- lounging on a velvet couch and staring at me with avid curiosity. This time she was not attended by slaves. The warriors chained me to the wall-every wall in that accursed palace seemed to have rings for captives-and left us alone.

I leaned back among the furs and pillows, finding their downy contact irksome to my iron-hard frame, unaccustomed to soft living of any kind, and for a wearisome time the Queen of Yugga surveyed me without speaking. Her eyes had a hypnotic quality; I distinctly felt their impact. But I felt too much like a chained beast on exhibition to be aware of any feeling but one of rising resentment. I fought it down. A burst of berserk fury might break the slender chains that held me, and rid the world of Yasmeena, but Altha and I would still be prisoners on that accursed rock from which legend said there was no escape save through the air.

"Who are you?" Yasmeena demanded abruptly. "I have seen men with skins smoother even than yours, but never a hairless white man before."

Before I could ask her where she had seen hairless men, if not among her own people, she continued: "Nor have I seen eyes like yours. They are like a deep cold lake, yet they blaze and smolder like the cold blue flame that dances forever above Xathar. What is your name? Whence come you? The girl Altha said you came out of the wilderness and dwelt in her city, defeating its mighty men in single combat. But she does not know from what land you came, she says. Speak, and do not lie."

"I'll speak but you'll think I lie," I grunted. "I am Esau Cairn, whom the men of Koth call Ironhand. I come from another world in another solar system. Chance, or the whim of a scientist whom you would call a magician, cast me on this planet. Chance again threw me among the Kothans. Chance carried me to Yugga. Now I have spoken. Believe me or not, as you will."

"I believe you," she answered. "Of old, men passed from star to star, There are beings now which traverse the cosmos. I would study you. You shall live-for a while, at least. But you must wear those chains, for I read the fury of the beast in your eyes, and know you would rend me if you could."

"What of Altha!" I asked.

"Well, what of her?" She seemed surprised at the question.

"What have you done with her?" I demanded.

"She will serve me with the rest, until she displeases me. Why do you speak of another woman, when you are talking to me? I am not pleased."

Her eyes began to glitter. I never saw eyes like Yasmeena's. They changed with every shift of mood and whim, and they mirrored passions and angers and desires beyond the maddest dreams of humanity.

"You do not blench," she said softly. "Man, do you know what it is for Yasmeena to be displeased? Then blood flows like water, Yugga rings with screams of agony, and the very gods hide their heads in horror."

The way she said it turned my blood cold, but the red anger of the primitive would not down. The feel of my strength came upon me, and I knew that I could tear that golden ring from the stone and rip out her life before she could leap from her couch, if it came to that. So I laughed, and my laughter thrummed with blood-lust. She started up and eyed me closely.

"Are you mad, to laugh?" she asked. "No, that was not mirth-it was the growl of a hunting leopard. It is in your mind to leap and kill me, but if you do, the girl Altha will suffer for your crime. Yet you interest me. No man has ever laughed at me before. You shall live-for a while." She clapped her hands and the warriors entered. "Take him back to his chamber," she directed. "Keep him chained there until I send for him again."

And so began my third captivity on Almuric, in the black citadel of Yugga, on the rock Yuthla, by the river of Yogh, in the land of Yagg.

CHAPTER 9

Much I learned of the ways of that terrible people, who have reigned over Almuric since ages beyond the memory of man. They might have been human once, long ago, but I doubt it. I believe they represented a separate branch on the tree of evolution, and that it is only an incredible freak of coincidence which cast them in a mold so similar to man, instead of the shapes of the abysmal, howling, blasphemous dwellers of Outer Darkness.

In many ways they seemed, superficially, human enough, but if one followed their lines of consciousness far enough, he would come upon phases inexplicable and alien to humanity. As far as pure intellect went, they were superior to the hairy Guras. But they lacked altogether the decency, honesty, courage, and general manliness of the ape-men. The Guras were quick to wrath, savage and brutal in their anger; but there was a studied cruelty about the Yagas which made the others seem like mere rough children. The Yagas were merciless in their calmest moments; roused to anger, their excesses were horrible to behold.

They were a numerous horde, the warriors alone numbering some twenty thousand. There were more women than men, and with their slaves, of which each male and female Yaga possessed a goodly number, the city of Yugga was fully occupied. Indeed, I was surprised to learn of the multitudes of people who dwelt there, considering the comparative smallness of the rock Yuthla on which the city was built. But its space was greater vertically than horizontally. The castles and towers soared high into the air, and several tiers of chambers and corridors were sunk into the rock itself. When the Yagas felt themselves crowded for space, they simply butchered their slaves. I saw no children; losses in war were comparatively slight, and plagues and diseases unknown. Children were produced only at regular intervals, some three centuries apart. The last flock had come of age; the next brood was somewhere in the dim distance of the future.

The lords of Yugga did no sort of work, but passed their lives in sensual pleasures. Their knowledge and adeptness at debauchery would have shamed the most voluptuous libertine in later Rome. Their debauches were interrupted only by raids on the outer world in order to procure women slaves.

The town at the foot of the cliff was called Akka, the blue people Akki, or Akkas. They had been subject to the Yagas as far back as tradition extended. They were merely stupid work-animals, laboring in the irrigated fields of fruits and edible plants, and otherwise doing the will of their masters, whom they considered superior beings, if not veritable gods. They worshipped Yasmeena as a deity. Outside of continual toil, they were not mistreated. Their women were ugly and beastlike. The winged people had a keen asthetic sense, though their interest in the beauty of the lower orders was sadistic and altogether beastly. The Akkas never came into the upper city, except when there was work to be done there, too heavy for the women slaves. Then they ascended and descended by means of great silken ladders let down from the rock. There was no road leading up from below, since the Yagas needed none. The cliffs could not be scaled; so the winged people had no fear of an Akka uprising.

The Yaga women were likewise prisoners on the rock Yuthla. Their wings were carefully removed at birth. Only the infants destined to become queens of Yugga were spared. This was done in order to keep the male sex in supremacy, and indeed, I was never able to learn how, and at what distant date, the men of Yugga gained supremacy over their women; for, judging from Yasmeena, the winged women were superior to their mates in agility, endurance, courage and even in strength. Clipping their wings kept them from developing their full superiority.

Yasmeena was an example of what a winged woman could be. She was taller than the other Yaga females, who in turn were taller than the Gura women, and though voluptuously shaped, the steel thews of a wildcat lurked in her slender rounded limbs. She was young-all the women of Yugga looked young. The average life-span of the Yaga was nine hundred years. Yasmeena had reigned over Yugga for four hundred years. Three winged princesses of royal blood had contested with her for the right to rule, and she had slain each of them, fighting with naked hands in the regal octagonal chamber. As long as she could defend her crown against young claimants, she would rule.

The lot of the slaves in Yugga was hideous. None ever knew when she would be dismembered for the cooking-pot, and the lives of all were tormented by the cruel whims of their masters and mistresses. Yugga was as like Hell as any place could be. I do not know what went on in the palaces of the nobles and warriors, but I do know what took place daily in the palace of the Queen. There was never a day or night that those dusky walls did not re-echo screams of agony and piteous wails for mercy, mingled with vindictive maledictions, or lascivious laughter.

I never became accustomed to it, hard as I was physically and mentally. I think the only thing that kept me from going mad was the feeling that I must keep my sanity in order to protect Altha if I could. That was precious little; I was chained in my chamber; where the Kothan girl was, I had not the slightest idea, except that she was somewhere in the palace of Yasmeena, where she was protected from the lust of the winged men, but not from the cruelty of her mistress.

81

In Yugga I heard sounds and saw sights not to be repeated-not even to be remembered in dreams. Men and women, the Yagas were open and candid in their evil. Their utter cynicism banished ordinary scruples of modesty and common decency. Their bestialities were naked, unhidden and shameless. They followed their desires with one another, and practiced their tortures on their wretched slaves with no attempt at concealment. Deeming themselves gods, they considered themselves above the considerations that guide ordinary humans. The women were more vicious than the men, if such a thing were possible. The refinements of their cruelties toward their trembling slaves cannot be even hinted at. They were versed in every art of torture, both mental and physical. But enough. I can but hint at what is unrepeatable.

Those days of captivity seem like a dim nightmare. I was not badly treated, personally. Each day I was escorted on a sort of promenade about the palace-something on the order of giving a confined animal exercise. I was always accompanied by seven or eight warriors armed to the teeth, and always wore my chains. Several times on these promenades I saw Altha, going about her duties, but she always averted her gaze and hurried by. I understood and made no attempt to speak to her. I had placed her in jeopardy already by speaking of her to Yasmeena. Better let the queen forget about her, if possible. Slaves were safest when the Queen of Yagg remembered them least.

Somewhere, somehow, I found in me power to throttle my red rage and blind fury. When my very brain reeled with the lust to break my chains and explode into a holocaust of slaughter, I held myself with iron grasp. And the fury ate inward into my soul, crystallizing my hate. So the days passed, until the night that Yasmeena again sent for me.

CHAPTER 10

Yasmeena cupped her chin in her slim hands and fixed her great dark eyes on me. We were alone in a chamber I had never entered before. It was night. I sat on a divan opposite her, my limbs unshackled. She had offered me temporary freedom if I would promise not to harm her, and to go back into shackles when she bade me. I had promised. I was never a clever man, but my hate had sharpened my wits. I was playing a game of my own.

"What are you thinking of, Esau Ironhand?" she asked.

"I'm thirsty," I answered.

She indicated a crystal vessel near at hand. "Drink a little of the golden wine-not much, or it will make you drunk. It is the most powerful drink in the world. Not even I can quaff that vessel without lying senseless for hours. And you are unaccustomed to it."

I sipped a little of it. It was indeed heady liquor.

Yasmeena stretched her limbs out on her couch, and asked: "Why do you hate me? Have I not treated you well?"

"I have not said that I hated you," I countered. "You are very beautiful. But you are cruel."

She shrugged her winged shoulders. "Cruel? I am a goddess. What have I to do with either cruelty or mercy? Those qualities are for men. Humanity exists for my pleasure. Does not all life emanate from me?"

"Your stupid Akkas may believe that," I replied; "but I know otherwise, and so do you."

She laughed, not offended. "Well, I may not be able to create life, but I can destroy life at will. I may not be a goddess, but you would find it difficult to convince these foolish wenches who serve me that I am not all-powerful. No, Ironhand; gods are only another name for *Power*. I am Power on this planet; so I am a goddess. What do your hairy friends, the Guras, worship?"

"They worship Thak; at least they acknowledge Thak as the creator and preserver. They have no regular ritual of worship, no temples, altars or priests. Thak is the Hairy One, the god in the form of man. He bellows in the tempest, and thunders in the hills with the voice of the lion. He loves brave men, and hates weaklings, but he neither harms nor aids. When a male child is born, he blows into it courage and strength; when a warrior dies, he ascends to Thak's abode, which is a land of celestial plains, river and mountains, swarming with game, and inhabited by the spirits of departed warriors, who hunt, fight and revel forever as they did in life."

She laughed. "Stupid pigs. Death is oblivion. We Yagas worship only our own bodies. And to our bodies we make rich sacrifices with the bodies of the foolish little people."

"Your rule cannot last forever," I was moved to remark.

"It *has* lasted since beyond the gray dawn of Time's beginning. On the dark rock Yuthla my people have brooded through ages uncountable. Before the cities of the Guras dotted the plains, we dwelt in the land of Yagg. We were always masters. As we rule the Guras, so we ruled the mysterious race which possessed the land before the Guras evolved from the ape: the race which reared their cities of marble whose ruins now affright the moon, and which perished in the night.

"Tales! I could tell you tales to blast your reason! I could tell you of races which appeared from the mist of mystery, moved across the world in restless waves, and vanished in the midst of oblivion. We of Yugga have watched them come and go, each in turn bending beneath the yoke of our godship. We have endured, not centuries or millenniums, but cycles.

"Why should not our rule endure forever? How shall these Gura-fools overcome us? You have seen how it is when my hawks swoop from the air in the night on the cities of the ape-man. How then shall they attack us in our eyrie? To reach the land of Yagg they must cross the Purple River, whose waters race too swiftly to be swum. Only at the Bridge of Rocks can it be crossed, and there keen-eyed guards watch night and day. Once, the Guras did try to attack us. The watchers brought word of their coming and the men of Yagg were prepared. In the midst of the desert they fell on the invaders and destroyed them by thirst and madness and arrows showering upon them from the skies.

"Suppose a horde should fight its way through the desert and reach the rock Yuthla? They have the river Yogh to cross, and when they have crossed it, in the teeth of the Akki spears, what then? They could not scale the cliffs. No; no foreign foe will ever set foot in Yugga. If, by the wildest whim of the gods, such a thing *should* come to pass"- her beautiful features became even more cruel and sinister-"rather than submit to conquest I would loose the *Ultimate Horror*, and perish in the ruins of my city," she whispered, more to herself than to me.

"What do you say?" I asked, not understanding.

"There are secrets beneath the velvet coverings of the darkest secrets," she said. "Tread not where the very gods tremble. I said nothing-you heard nothing. Remember that!"

There was silence for a space, and then I asked a question I had long mulled over: "Whence come these red girls and yellow girls among your slaves?"

"You have looked southward from the highest towers on clear days, and seen a faint blue line rimming the sky far away? That is the Girdle that bands the world. Beyond that Girdle dwell the races from which come those alien slaves. We raid across the Girdle just as we raid the Guras, though less frequently."

I was about to ask more concerning these unknown races, when a timid tap came on the outer door. Yasmeena, frowning at the interruption, called a sharp question, and a frightened feminine voice informed her that the lord Gotrah desired audience. Yasmeena spat an oath at her, and bade her tell the lord Gotrah to go to the devil.

"No, I must see the fellow," she said rising. "Theta! Oh, Theta! Where has the little minx gone? I must do my own biddings, must I? Her buttocks shall smart for her insolence. Wait here, Ironhand. I'll see to Gotrah."

She crossed the cushion-strewn chamber with her lithe, long stride, and passed through the door. As it closed behind her, I was struck by what was nothing less than an inspiration. No especial reason occurred to me to urge me to feign drunkenness. It was intuition or blind chance that prompted me. Snatching up the crystal jug which contained the golden wine, I emptied it into a great golden vessel which stood half hidden beneath the fringe of a tapestry. I had drunk enough for the scent to be on my breath.

Then, as I heard footsteps and voices without, I extended myself quickly on a divan, the jug lying on its side near my outstretched hand. I heard the door open, and there was an instant's silence so intense as to be almost tangible. Then Yasmeena spat like an angry cat. "By the gods, he's emptied the jug? See how he lies in brutish slumber! Faugh! The noblest figure is abominable when besotted. Well, let us to our task. We need not fear to be overheard by him."

"Had I not better summon the guard and have him dragged to his cell?" came Gotrah's voice. "We cannot afford to take chances with this secret, which none has ever known except the Queen of Yugga and her major-domo."

I sensed that they came and stood over me, looking down. I moved vaguely and mumbled thickly, as if in drunken dreams.

Yasmeena laughed.

"No fear. He will know nothing before dawn. Yuthla could split and fall into Yogh without breaking his sottish dreams. The fool! This night he would have been lord of the world, for I would have made him lord of the Queen of the world-for one night. But the lion changes not his mane, nor the barbarian his brutishness."

"Why not put him to the torture?" grunted Gotrah.

"Because I want a man, not a broken travesty. Besides, his is a spirit not to be conquered by fire or steel. No. I am Yasmeena and I will make him love me before I feed him to the vultures. Have you placed the Kothan Altha among the Virgins of the Moon?"

"Aye, Queen of the dusky stars. A month and a half from this night she dances the dance of the Moon with the other wenches."

"Good. Keep them guarded day and night. If this tiger learns of our plans for his sweetheart, chains and bolts will not hold him."

"A hundred and fifty men guard the virgins," answered Gotrah. "Not even the Ironhand could prevail against them."

"It is well. Now to this other matter. Have you the parchment?"

"Aye."

"Then I will sign it. Give me the stylus."

I heard the crackle of papyrus and the scratch of a keen point, and then the Queen said:

"Take it now, and lay it on the altar in the usual place. As I promise in the writing, I will appear in the flesh tomorrow night to my faithful subjects and worshippers, the blue pigs of Akka. Ha! ha! ha! I never fail to be amused at the animal-like awe on their stupid countenances when I emerge from the shadows of the golden screen, and spread my arms above them in blessing. What fools they are, not in all these ages, to have discovered the secret door and the shaft that leads from their temple to this chamber."

"Not so strange," grunted Gotrah. "None but the priest ever comes into the temple except by special summons, and he is far too superstitious to go meddling behind the screen. Anyway, there is no sign to mark the secret door from without."

"Very well," answered Yasmeena. "Go."

I heard Gotrah fumbling at something, then a slight grating sound. Consumed by curiosity, I dared open one eye a slit, in time to glimpse Gotrah disappearing through a black opening that gaped in the middle of the stone floor, and which closed after him. I quickly shut my eye again and lay still, listening to Yasmeena's quick pantherish tread back and forth across the floor.

Once she came and stood over me. I felt her burning gaze and heard her curse beneath her breath. Then she struck me viciously across the face with some kind of jeweled ornament that tore my skin and started a trickle of blood. But I lay without twitching a muscle, and presently she turned and left the chamber, muttering.

As the door closed behind her I rose quickly, scanning the floor for some sign of the opening through which Gotrah had gone. A furry rug had been drawn aside from the center of the floor, but in the polished black stone I searched in vain for a crevice to denote the hidden trap. I momentarily expected the return of Yasmeena, and my heart pounded within me. Suddenly, under my very hand, a section of the floor detached itself and began to move upward. A pantherish bound carried me behind a tapestried couch, where I crouched, watching the trap rise upward. The narrow head of Gotrah appeared, then his winged shoulders and body.

He climbed up into the chamber, and as he turned to lower the lifted trap, I left the floor with a catlike leap that carried me over the couch and full on his shoulders.

He went down under my weight, and my gripping fingers crushed the yell in his throat. With a convulsive heave he twisted under me, and stark horror flooded his face as he glared up at me. He was down on the cushioned stone, pinned under my iron bulk. He clawed for the dagger at his girdle, but my knee pinned it down. And crouching on him, I gutted my mad hate for his cursed race. I strangled him slowly, gloatingly, avidly watching his features contort and his eyes glaze. He must have been dead for some minutes before I loosed my hold.

Rising, I gazed through the open trap. The light from the torches of the chamber shone down a narrow shaft, into which was cut a series of narrow steps, that evidently led down into the bowels of the rock Yuthla. From the conversation I had heard, it must lead to the temple of the Akkas, in the town below. Surely I would find Akka no harder to escape from than Yugga. Yet I hesitated, my heart torn at the thought of leaving Altha alone in Yugga. But there was no other way. I did not know in what part of that devil-city she was imprisoned, and I remembered what Gotrah had said of the great band of warriors guarding her and the other virgins.

Virgins of the Moon! Cold sweat broke out on me as the full significance of the phrase became apparent. Just what the festival of the Moon was I did not fully know, but I had heard hints and scattered comments among the Yaga women, and I knew it was a beastly saturnalia, in which the full frenzy of erotic ecstasy was reached in the dying gasps of the wretches sacrificed to the only god the winged people recognized-their own inhuman lust.

The thought of Altha being subjected to such a fate drove me into a berserk frenzy, and steeled my resolution. There was but one chance- to escape myself, and try to reach Koth and bring back enough men to attempt a rescue. My heart sank as I contemplated the difficulties in the way, but there was nothing else to be done.

Lifting Gotrah's limp body I dragged it out of the chamber through a door different from that through which Yasmeena had gone; and traversing a corridor without meeting anyone, I concealed the corpse behind some tapestries. I was certain that it would be found, but perhaps not until I had a good start. Perhaps its presence in another room than the chamber of the trap might divert suspicion from my actual means of escape, and lead Yasmeena to think that I was merely hiding somewhere in Yugga.

But I was crowding my luck. I could not long hope to avoid detection if I lingered. Returning to the chamber, I entered the shaft, lowering the trap above me. It was pitch-dark, then, but my groping fingers found the catch that worked the trap, and I felt that I could return if I found my way blocked below. Down those inky stairs I groped, with an uneasy feeling that I might fall into some pit or meet with some grisly denizen of the underworld. But nothing occurred, and at last the steps ceased and I groped my way along a short corridor that ended at a blank wall. My fingers encountered a metal catch, and I shot the bolt, feeling a section of the wall revolving under my hands. I was dazzled by a dim yet lurid light, and blinking, gazed out with some trepidation.

I was looking into a lofty chamber that was undoubtedly a shrine. My view was limited by a large screen of carved gold directly in front of me, the edges of which flamed dully in the weird light.

Gliding from the secret door, I peered around the screen. I saw a broad room, made with the same stern simplicity and awesome massiveness that characterized Almuric architecture. The ceiling was lost in the brooding shadows; the walls were black, dully gleaming, and unadorned. The shrine was empty except for a block of ebon stone, evidently an altar, on which blazed the lurid flame I had noted, and which seemed to emanate from a great somber jewel set upon the altar. I noticed darkly stained channels on the sides of that altar, and on the dusky stone lay a roll of white parchment-Yasmeena's word to her worshippers. I had stumbled into the Akka holy of holies-uncovered the very root and base on which the whole structure of Akka theology was based: the supernatural appearances of revelations from the goddess, and the appearance of the goddess herself in the temple. Strange that a whole religion should be based on the ignorance of the devotees concerning a subterranean stair! Stranger still, to an Earthly mind, that only the lowest form of humanity on Almuric should possess a systematic and ritualistic religion, which Earth people regard as sure token of the highest races!

But the cult of the Akkas was dark and weird. The whole atmosphere of the shrine was one of mystery and brooding horror. I could imagine the awe of the blue worshippers to see the winged goddess emerging from behind the golden screen, like a deity incarnated from cosmic emptiness.

Closing the door behind me, I glided stealthily across the temple. Just within the door a stocky blue man in a fantastic robe lay snoring lustily on the naked stone. Presumably he had slept tranquilly through Gotrah's ghostly visit. I stepped over him as gingerly as a cat treading wet earth, Gotrah's dagger in my hand, but he did not awaken. An instant later I stood outside, breathing deep of the river-laden night air.

The temple lay in the shadow of the great cliffs. There was no moon, only the myriad millions of stars that glimmer in the skies of Almuric. I saw no lights anywhere in the village, no movement. The sluggish Akkas slept soundly.

Stealthily as a phantom I stole through the narrow streets, hugging close to the sides of the squat stone huts. I saw no human until I reached the wall. The drawbridge that spanned the river was drawn up, and just within the gate sat a blue man, nodding over his spear. The senses of the Akkas were dull as those of any beasts of burden. I could have knifed the drowsy watchman where he sat, but I saw no need of useless murder. He did not hear me, though I passed within forty feet of him. Silently I glided over the wall, and silently I slipped into the water.

Striking out strongly, I forged across the easy current, and reached the farther bank. There I paused only long enough to drink deep of the cold river water; then I struck out across the shadowed desert at a swinging trot that eats up miles-the gait with which the Apaches of my native Southwest can wear out a horse.

In the darkness before dawn I came to the banks of the Purple River, skirting wide to avoid the watchtower which jutted dimly against the star-flecked sky. As I crouched on the steep bank and gazed down into the rushing swirling current, my heart sank. I knew that, in my fatigued condition, it was madness to plunge into the maelstrom. The strongest swimmer that either Earth or Almuric ever bred had been helpless among those eddies and whirlpools. There was but one thing to be done-try to reach the Bridge of Rocks before dawn broke, and take the desperate chance of slipping across under the eyes of the watchers. That, too, was madness, but I had no choice.

But dawn began to whiten the desert before I was within a thousand yards of the Bridge. And looking at the tower, which seemed to swim slowly into clearer outline, etched against the dim sky, I saw a shape soar up from the turrets and wing its way toward me. I had been discovered. Instantly, a desperate plan occurred to me. I began to stagger erratically, ran a few paces, and sank down in the sand near the river bank. I heard the beat of wings above me as the suspicious harpy circled; then I knew he was dropping earthward. He must have been on solitary sentry duty, and had come to investigate the matter of a lone wanderer, without waking his mates.

Watching through slitted lids, I saw him strike the earth near by, and walk about me suspiciously, scimitar in hand. At last he pushed me with his foot, as if to find if I lived. Instantly my arm hooked about his legs, bringing him down on top of me. A single cry burst from his lips, half-stifled as my fingers found his throat; then in a great heaving and fluttering of wings and lashing of limbs, I heaved him over and under me. His scimitar was useless at such close quarters. I twisted his arm until his numbed fingers slipped from the hilt; then I choked him into submission. Before he regained his full faculties, I bound his wrists in front of him with his girdle, dragged him to his feet, and perched myself astride his back, my legs locked about his torso. My left arm was hooked about his neck, my right hand pricked his hide with Gotrah's dagger.

In a few low words I told him what he must do, if he wished to live. It was not the nature of a Yaga to sacrifice himself, even for the welfare of his race. Through the rose-pink glow of dawn we soared into the sky, swept over the rushing Purple River, and vanished from the sight of the land of Yagg, into the blue mazes of the northwest.

90

CHAPTER 11

I drove that winged devil unmercifully. Not until sunset did I allow him to drop earthward. Then I bound his feet and wings so he could not escape, and gathered fruit and nuts for our meal. I fed him as well as I fed myself. He needed strength for the flight. That night the beasts of prey roared perilously close to us, and my captive turned ashy with fright, for we had no way of making a protecting fire, but none attacked us. We had left the forest of the Purple River far, far behind, and were among the grasslands. I was taking the most direct route to Koth, led by the unerring instinct of the wild. I continually scanned the skies behind me for some sign of pursuit, but no winged shapes darkened the southern horizon.

It was on the fourth day that I spied a dark moving mass in the plains below, which I believed was an army of men marching. I ordered the Yaga to fly over them. I knew that I had reached the vicinity of the wide territory dominated by the city of Koth, and there was a chance that these might be men of Koth. If so, they were in force, for as we approached I saw there were several thousand men, marching in some order.

So intense was my interest that it almost proved my undoing. During the day I left the Yaga's legs unbound, as he swore that he could not fly otherwise, but I kept his wrists bound. In my engrossment I did not notice him furtively gnawing at the thong. My dagger was in its sheath, since he had shown no recent sign of rebellion. My first intimation of revolt was when he wheeled suddenly sidewise, so that I lurched and almost lost my grip on him. His long arm curled about my torso and tore at my girdle, and the next instant my own dagger gleamed in his hand.

There ensued one of the most desperate struggles in which I have ever participated. My near fall had swung me around, so that instead of being on his back, I was in front of him, maintaining my position only by one hand clutching his hair, and one knee crooked about his leg. My other hand was locked on his dagger wrist, and there we tore and twisted, a thousand feet in the air, he to break away and let me fall to my death, or to drive home the dagger in my breast, I to maintain my grip and fend off the gleaming blade.

On the ground my superior weight and strength would quickly have settled the issue, but in the air he had the advantage. His free hand beat and tore at my face, while his unimprisoned knee drove viciously again and again for my groin. I hung grimly on, taking the punishment without flinching, seeing that our struggles were dragging us lower and lower toward the earth.

Realizing this, he made a final desperate effort. Shifting the dagger to his free hand, he stabbed furiously at my throat. At the same instant I gave his head a terrific downward wrench. The impetus of both our exertions whirled us down and over, and his stroke, thrown out of line by our erratic convulsion, missed its mark and sheathed the dagger in his own thigh. A terrible cry burst from his lips, his grasp went limp as he half fainted from the pain and shock, and we rushed plummet-like earthward. I strove to turn him beneath me, and even as I did, we struck the earth with a terrific concussion.

From that impact I reeled up dizzily. The Yaga did not move; his body had cushioned mine, and half the bones in his frame must have been splintered.

A clamor of voices rang on my ears, and turning, I saw a horde of hairy figures rushing toward me. I heard my own name bellowed by a thousand tongues. I had found the men of Koth.

A hairy giant was alternately pumping my hand and beating me on the back with blows that would have staggered a horse, while bellowing: "Ironhand! By Thak's jawbones, *Ironhand*! Grip my hand, old war-dog! Hell's thunders, I've known no such joyful hour since the day I broke old Khush of Tanga's back!"

There was old Khossuth Skullsplitter, somber as ever, Thab the Swift, Gutchluk Tigerwrath-nearly all the mighty men of Koth. And the way they smote my back and roared their welcome warmed my heart as it was never warmed on Earth, for I knew there was no room for insincerity in their great simple hearts.

"Where have you been, Ironhand?" exclaimed Thab the Swift. "We found your broken carbine out on the plains, and a Yaga lying near it with his skull smashed; so we concluded that you had been done away with by those winged devils. But we never found your body-and now you come tumbling through the skies locked in combat with another flying fiend! Say, have you been to Yugga?" He laughed as a man laughs when he speaks a jest.

"Aye to Yugga, on the rock Yuthla, by the river Yogh, in the land of Yagg," I answered. "Where is Zal the Thrower?"

"He guards the city with the thousand we left behind," answered Khossuth.

"His daughter languishes in the Black City," I said. "On the night of the full moon, Altha, Zal's daughter, dies with five hundred other girls of the Guras-unless we prevent it."

A murmur of wrath and horror swept along the ranks. I glanced over the savage array. There were a good four thousand of them; no bows were in evidence, but each man bore his carbine. That meant war, and their numbers proved it was no minor raid.

"Where are you going?" I asked.

"The men of Khor move against us, five thousand strong," answered Khossuth. "It is the death grapple of the tribes. We march to meet them afar off from our walls, and spare our women the horrors of the war."

"Forget the men of Khor!" I cried passionately. "You would spare the feelings of your women-yet thousands of your women suffer the tortures of the damned on the ebon rock of Yuthla! Follow me! I will lead you to the stronghold of the devils who have harried Almuric for a thousand ages!"

"How many warriors?" asked Khossuth uncertainly.

"Twenty thousand."

A groan rose from the listeners.

"What could our handful do against that horde?"

"I'll show you!" I exclaimed. "I'll lead you into the heart of their citadel!"

"Hai!" roared Ghor the Bear, brandishing his broadsword, always quick to take fire from my suggestions. "That's the word! Come on, sir brothers! Follow Ironhand! He'll show us the way!"

"But what of the men of Khor?" expostulated Khossuth. "They are marching to attack us. We must meet them."

Ghor grunted explosively as the truth of this assertion came home to him and all eyes turned toward me.

"Leave them to me," I proposed desperately. "Let me talk with them-"

"They'll hack off your head before you can open your mouth," grunted Khossuth.

"That's right," admitted Ghor. "We've been fighting the men of Khor for fifty thousand years. Don't trust them, comrade."

"I'll take the chance," I answered.

"The chance you shall have, then," said Gutchluk grimly. "For there they come!" In the distance we saw a dark moving mass.

"Carbines ready!" barked old Khossuth, his cold eyes gleaming. "Loosen your blades, and follow me."

"Will you join battle tonight?" I asked. He glanced at the sun. "No. We'll march to meet them, and pitch camp just out of gunshot. Then with dawn we'll rush them and cut their throats."

"They'll have the same idea," explained Thab. "Oh, it will be great fun!"

"And while you revel in senseless bloodshed," I answered bitterly, "your daughters and theirs will be screaming vainly under the tortures of the winged people over the river Yogh. Fools! Oh, you fools!"

"But what can we do?" expostulated Gutchluk.

"Follow me!" I yelled passionately. "We'll march to meet them, and I'll go on to them alone."

I wheeled and strode across the plain, and the hairy men of Koth fell in behind me, with many headshakes and mutterings. I saw the oncoming mass, first as a mingled blur; then the details stood out- hairy bodies, fierce faces, gleaming weapons-but I swung on heedlessly. I knew neither fear nor caution; my whole being seemed on fire with the urgency of my need and desire.

Several hundred yards separated the two hosts when I dashed down my single weapon-the Yaga dagger-and shaking off Ghor's protesting hands, advanced alone and unarmed, my hands in the air; palms toward the enemy.

These had halted, drawn up ready for action. The unusualness of my actions and appearance puzzled them. I momentarily expected the crack of a carbine, but nothing happened until I was within a few yards of the foremost group, the mightiest men clustered about a tall figure that was their chief-old Bragi, Khossuth had told me. I had heard of him, a hard, cruel man, moody and fanatical in his hatreds.

"Stand!" he shouted, lifting his sword. "What trick is this? Who are you who comes with empty hands in the teeth of war?"

"I am Esau Ironhand, of the tribe of Koth," I answered. "I would parley with you."

"What madman is this?" growled Bragi. "Than-a bullet through his head."

But the man called Than, who had been staring eagerly at me, gave a shout instead and threw down his carbine.

"Not if I live!" he exclaimed, advancing toward me his arms outstretched. "By Thak, it is he! Do you not remember me, Than Swordswinger, whose life you saved in the hills?"

He lifted his chin to display a great scar on his corded neck.

"You are he who fought the saber-tooth! I had not dreamed you survived those awful wounds."

"We men of Khor are hard to kill!" he laughed joyously, throwing his arms about me in a bear-like embrace. "What are you doing among the dogs of Koth? You should be fighting with us!"

"If I have my way there will be no fighting," I answered. "I wish only to talk with your chiefs and warriors. There is nothing out of the way about that."

"True!" agreed Than Swordswinger. "Bragi, you will not refuse him this?"

Bragi growled in his beard, glaring at me.

"Let your warriors advance to that spot." I indicated the place I meant. "Khossuth's men will come up on the other side. There both hordes will listen to what I have to say. Then, if no agreement can be reached, each side shall withdraw five hundred yards and after that follow its own initiative."

"You are mad!" Old Bragi jerked his beard with a shaking hand of rage. "It is treachery. Back to your kennel, dog!"

"I am your hostage," I answered. "I am unarmed. I will not move out of your sword reach. If there is treachery, strike me down on the spot."

"But why?"

"I have been captive among the Yagas!" I exclaimed. "I have come to tell the Guras what things occur in the land of Yagg!"

"The Yagas took my daughter!" exclaimed a warrior, pushing through the ranks. "Did you see her in Yagg?"

"They took my sister!"-"And my young bride"-"And my niece!" shouts rose in chorus, as men swarmed about me, forgetful of their enemies, shaking me in the intensity of their feeling.

"Back, you fools!" roared Bragi, smiting with the flat of his sword. "Will you break your ranks and let the Kothans cut you down? Do you not see it is a trick?"

"It is no trick!" I cried. "Only listen to me, in God's name!"

They swept away Bragi's protests. There was a milling and stamping, during which only a kindly Providence kept the nerve-taut Kothans from pouring a volley into the surging mass of their enemies, and presently a sort of order was evolved. A shouted conference finally resulted in approximately the position I had asked for-a semicircle of Khorans over against a similar formation composed of Kothans. The close proximity almost caused the tribal wrath to boil over. Jaws jutted, eyes blazed, hairy hands clutched convulsively at carbine stocks. Like wild dogs those wild men glared at each other, and I hastened to begin my say.

I was never much of a talker, and as I strode between those hostile hordes I felt my fire die out in cold ague of helplessness. A million ages of traditional war and feud rose up to confound me. One man against the accumulated ideas, inhibitions, and customs of a whole world, built up through countless millenniums-the thought crushed and paralyzed me. Then blind rage swept me at the memory of the horrors of Yugga, and the fire blazed up again and enveloped the world and made it small, and on the wings of that conflagration I was borne to heights of which I had never dreamed.

No need for fiery oratory to tell the tale I had to tell. I told it in the plainest, bluntest language possible, and the knowledge and feeling that lay behind the telling made those naked words pulse, and burn like acid.

I told of the hell that was Yugga. I told of young girls dying beneath the excesses of black demons-of women lashed to gory ribbons, mangled on the wheel, sundered on the rack, flayed alive, dismembered alive-of the torments that left the body unharmed, but sucked the mind empty of reason and left the victim a blind, mewing imbecile. I told them-oh God, I cannot repeat all I told them, at the memory of which I am even now sickened almost unto death.

Before I had finished, men were bellowing and beating their breasts with

96

their clenched fists, and weeping in agony of grief and fury.

I lashed them with a last whip of scorpions. "These are your women, your own flesh and blood, who scream on the racks of Yugga! You call yourselves men-you strut and boast and swagger, while these winged devils mock you. Men! Ha!" I laughed as a wolf barks, from the depths of my bitter rage, and agony. "Men! Go home and don the skirts of women!"

A terrible yell arose. Clenched fists were brandished, bloodshot eyes flamed at me, hairy throats bayed their anguished fury. "You lie, you dog! Damn you, you lie! We *are* men! Lead us against these devils or we will rend you!"

"If you follow me," I yelled, "few of you will return. You will suffer and you will die in hordes. But if you had seen what I have seen, you would not wish to live. Soon approaches the time when the Yagas will clean their house. They are weary of their slaves. They will destroy those they have, and fare forth into the world for more. I have told you of the destruction of Thugra. So it will be with Khor; so it will be with Koth-when winged devils swoop out of the night. Follow me to Yugga-I will show you the way. If you are men, follow me!"

Blood burst from my lips in the intensity of my appeal, and as I reeled back, in a state of complete collapse from overwrought nerves and strain, Ghor caught me in his mighty arms.

Khossuth rose like a gaunt ghost. His ghostly voice soared out across the tumult.

"I will follow Esau Ironhand to Yugga, if the men of Khor will agree to a truce until our return. What is your answer, Bragi?"

"No!" roared Bragi. "There can be no peace between Khor and Koth. The women in Yugga are lost. Who can war against demons? Up, men, back to your place! No man can twist me with mad words to forget old hates."

He lifted his sword, and Than Swordswinger, tears of grief and fury running down his face, jerked out his poniard and drove it to the hilt in the heart of his king. Wheeling to the bewildered horde, brandishing the bloody dagger, his body shaken with sobs of frenzy, he yelled:

"So die all who would make us traitors to our own women! Draw your swords, all men of Khor who will follow me to Yugga!"

Five thousand swords flamed in the sun, and a deep-throated thunderous roar shook the very sky. Then wheeling to me, his eyes coals of madness:

"Lead us to Yugga, Esau Ironhand!" cried Than Swordswinger. "Lead us to Yagg, or lead us to Hell! We will stain the waters of Yogh with blood, and the Yagas will speak of us with shudders for ten thousand times a thousand years!"

Again the clangor of swords and the roar of frenzied men maddened the sky.

CHAPTER 12

Runners were sent to the cities, to give word of what went forward. Southward we marched, four thousand men of Koth, five thousand of Khor. We moved in separate columns, for I deemed it wise to keep the tribes apart until the sight of their oppressors should again drown tribal feelings.

Our pace was much swifter than that of an equal body of Earth soldiers. We had no supply trains. We lived off the land through which we passed. Each man bore his own armament-carbine, sword, dagger, canteen, and ammunition pouch. But I chafed at every mile. Sailing through the air on the back of a captive Yaga had spoiled me for marching. It took us days to cover ground the flying men had passed over in hours. Yet we progressed, and some three weeks from the time we began the march, we entered the forest beyond which lay the Purple River and the desert that borders the land of Yagg.

We had seen no Yagas, but we went cautiously now. Leaving the bulk of our force encamped deep in the forest, I went forward with thirty men, timing our march so that we reached the bank of the Purple River a short time after midnight, just before the setting of the Moon. My purpose was to find a way to prevent the tower guard from carrying the news of our coming to Yugga, so that we might cross the desert without being attacked in the open, where the numbers and tactics of the Yagas would weigh most heavily against us.

Khossuth suggested that we lie in wait among the trees along the bank, and pick the watchers off at long range at dawn, but this I knew to be impossible. There was no cover along the water's edge, and the river lay between. The men in the tower were out of our range. We might creep near enough to pick off one or two, but it was imperative that all should perish, since the escape of one would be enough to ruin our plans.

So we stole through the woods until we reached a point a mile upstream, opposite a jutting tongue of rock, toward which, I believed, a current set in from the center of the stream. There we placed in the water a heavy, strong catamaran we had constructed, with a long powerful rope. I got upon the craft with four of the best marksmen of the combined horde-Thab the Swift, Skel the Hawk, and two warriors of Khor. Each of us bore two carbines, strapped to our backs.

We bent to work with crude oars, though our efforts seemed ludicrously futile in the teeth of that flood. But the raft was long enough and heavy enough not to be spun by every whirlpool we crossed, and by dint of

99

herculean effort we worked out toward the middle of the stream. The men on shore paid out the rope, and it acted as a sort of brace, swinging us around in a wide arc that would have eventually brought us back to the bank we had left, had not the current we hoped for suddenly caught us and hurled us at dizzy speed toward the projecting tongue of rock. The raft reeled and pitched, driving its nose under repeatedly, until sometimes we were fully submerged. But our ammunition was waterproof, and we had lashed ourselves to the logs; so we hung on like drowned rats, until our craft was dashed against the rocky point.

It hung there for a breathless instant, in which time it was touch and go. We slashed ourselves loose, jumped into the water which swirled arm-pit deep about us, and fought our way along the point, clinging tooth and nail to every niche or projection, while the foaming current threatened momentarily to tear us away and send us after our raft which had slid off the ledge and was dancing away down the river.

We did make it, though, and hauled ourselves upon the shore at last, half dead from buffeting and exhaustion But we could not stop to rest, for the most delicate part of our scheme was before us. It was necessary that we should not be discovered before dawn gave us light enough to see the sights of our carbines, for the best marksman in the world is erratic by starlight. But I trusted to the chance that the Yagas would be watching the river, and paying scant heed to the desert behind them.

So in the darkness that precedes dawn, we stole around in a wide circle, and the first hint of light found us lying in a depression we had scraped in the sand not over four hundred yards to the south of the tower.

It was tense waiting, while the dawn lifted slowly over the land, and objects became more and more distinct. The roar of the water over the Bridge of Rocks reached us plainly, and at last we were aware of another sound. The clash of steel reached us faintly through the water tumult. Ghor and others were advancing to the river bank, according to my instructions. We could not see any Yagas on the tower; only hints of movement along the turrets. But suddenly one whirled up into the morning sky and started south at headlong speed. Skel's carbine cracked and the winged man, with a loud cry, pitched sideways and tumbled to earth.

There followed an instant of silence; then five winged shapes darted into the air, soaring high. The Yagas sensed what was occurring, and were chancing all on a desperate rush, hoping that at least one might get through. We all fired, but I scored a complete miss, and Thab only slightly wounded his man. But the others brought down the man I had missed, while Thab's second shot dropped the wounded Yaga. We reloaded hastily, but no more came from the tower. Six men watched there, Yasmeena had said. She had spoken the truth.

We cast the bodies into the river. I crossed the Bridge of Rocks, leaping from boulder to boulder, and told Ghor to take his men back into the forest, and to bring up the host. They were to camp just within the fringe of woods, out of sight from the sky. I did not intend to start across the desert until nightfall.

Then I returned to the tower and attempted to gain entrance, but found no doors, only a few small barred windows. The Yagas had entered it from the top. It was too tall and smooth to be climbed, so we did the only thing left to do. We dug pits in the sand and covered them with branches, over which we scattered dust. In these pits we concealed our best marksmen, who lay all day, patiently scanning the sky. Only one Yaga came winging across the desert. No human was in sight, and he was not suspicious until he poised directly over the tower. Then, when he saw no watchmen, he became alarmed, but before he could race away, the reports of half a dozen carbines brought him tumbling to the earth in a whirl of limbs and wings.

As the sun sank, we brought the warriors across the Bridge of Rocks, an accomplishment which required some time. But at last they all stood on the Yaga side of the river, and with our canteens well filled, we started at quick pace across the narrow desert. Before dawn we were within striking distance of the river.

Having crossed the desert under cover of darkness, I was not surprised that we were able to approach the river without being discovered. If any had been watching from the citadel, alert for anything suspicious, they would have discerned our dark mass moving across the sands under the dim starlight. But I knew that in Yugga no such watch was ever kept, secure as the winged people felt in the protection of the Purple River, of the watchmen in the tower, and of the fact that for centuries no Gura raid had dared the bloody doom of former invaders. Nights were spent in frenzied debauchery, followed by sodden sleep. As for the men of Akka, these slow-witted drudges were too habitually drowsy to constitute much menace against our approach, though I knew that once roused they would fight like animals.

So three hundred yards from the river we halted, and eight thousand men under Khossuth took cover in the irrigation ditches that traversed the fields of fruit. The waving fronds of the squat trees likewise aided in their concealment. This was done in almost complete silence. Far above us towered the somber rock Yuthla. A faint breeze sprang up, forerunner of dawn. I led the remaining thousand warriors toward the river bank. Halting them a short distance from it, I wriggled forward on my belly until my hands were at the water's edge. I thanked the Fates that had given me such men to lead. Where civilized men would have floundered and blundered, the Guras moved as easily and noiselessly as stalking panthers.

Across from me rose the wall, sheer from the steep bank, that guarded Akka. It would be hard to climb in the teeth of spears. At the first crack of dawn, the bridge, which towered gauntly against the stars, would be lowered so that Akkas might go into the fields to work. But before then the rising light would betray our forces.

With a word to Ghor, who lay at my side, I slid into the water and struck out for the farther shore, he following. Reaching a point directly below the bridge, we hung in the water, clutching the slippery wall, and looked about for some way of climbing it. There the water, near the bank, was almost as deep as in midstream. At last Ghor found a crevice in the masonry, wide enough to give him a grip for his hands. Then bracing himself, he held fast while I clambered on his shoulders. Standing thus I managed to reach the lower part of the lifted bridge, and an instant later I drew myself up. The erected bridge closed the gap in the wall. I had to clamber over the barrier. One leg was across, when a figure sprang out of the shadows, yelling a warning. The watchman had not been as drowsy as I had expected.

He leaped at me, the starlight glinting on his spear. With a desperate twist of my body, I avoided the whistling blade, though the effort almost toppled me from the wall. My out-thrown hand gripped his lank hair as he fell against the coping with the fury of his wasted thrust, and jerking myself back into balance, I dealt him a crushing buffet on the ear with my clenched fist. He crumpled, and the next instant I was over the wall.

Ghor was bellowing like a bull in the river, mad to know what was taking place above him, and in the dim light the Akkas were swarming like bees out of their stony hives. Leaning over the barrier I stretched Ghor the shaft of the watchman's spear, and he came heaving and scrambling up beside me. The Akkas had stared stupidly for an instant; then realizing they were being invaded, they rushed, howling madly.

As Ghor sprang to meet them, I leaped to the great windlass that controlled the bridge. I heard the Bear's thunderous war cry boom above the squalling of the Akkas, the strident clash of steel and the crunch of splintered bone. But I had no time to look; it was taking all my strength to work the windlass. I had seen five Akkas toiling together at it; yet in the stress of the moment I accomplished its lowering single-handed, though sweat burst out on my forehead and my muscles trembled with the effort. But down it came, and the farther end touched the other bank in time to accommodate the feet of the warriors who sprang up and rushed for it.

I wheeled to aid Ghor, whose panting gasps I still heard amidst the clamor

102

of the melee. I knew the din in the lower town would soon rouse the Yagas and it was imperative that we gain a foothold in Akka before the shafts of the winged men began to rain among us.

Ghor was hard pressed when I turned from the bridge-head. Half a dozen corpses lay under his feet, and he wielded his great sword with a berserk lustiness that sheared through flesh and bone like butter, but he was streaming blood, and the Akkas were closing in on him.

I had no weapon but Gotrah's dagger, but I sprang into the fray and ripped a sword from the sinking hand of one whose heart my slim blade found. It was a crude weapon, such as the Akkas forge, but it had edge and weight, and swinging it like a club, I wrought havoc among the swarming blue men. Ghor greeted my arrival with a gasping roar of pleasure, and redoubled the fury of his tremendous strokes, so that the dazed Akkas momentarily gave back.

And in that fleeting interval, the first of the Guras swarmed across the bridge. In an instant fifty men had joined us. But there the matter was deadlocked. Swarm after swarm of blue men rushed from their huts to fall on us with reckless fury. One Gura was a match for three or four Akkas, but they swamped us by numbers. They crushed us back into the bridge mouth, and strive as we could, we could not advance enough to clear the way for the hundreds of warriors behind us who yelled and struggled to come to sword-strokes with the enemy. The Akkas pressed in on us in a great crescent, almost crushing us against the men behind us. They lined the walls, yelling and screaming and brandishing their weapons. There were no bows or missiles among them; their winged masters were careful to keep such things out of their hands.

In the midst of the carnage dawn broke, and the struggling hordes saw their enemies. Above us, I knew, the Yagas would be stirring. Indeed I thought I could already hear the thrash of wings above the roar of battle, but I could not look up. Breast to breast we were locked with the heaving, grunting hordes, so closely there was no room for sword-strokes. Their teeth and filthy nails tore at us beastlike; their repulsive body odor was in our nostrils. In the crush we writhed and cursed, each man striving to free a hand to strike.

My flesh crawled in dread of the arrows I knew must soon be raining from above, and even with the thought the first volley came like a whistling sheet of sleet. At my side and behind me men cried out, clutching at the feathered ends protruding from their bodies. But then the men on the bridge and on the farther bank, who had held their fire for fear of hitting their comrades in the uncertain light, began loosing their carbines at the Akkas. At that range their fire was devastating. The first volley cleared the wall, and climbing on the bridge rails the carbineers poured a withering fusillade over our heads into the close-massed horde that barred our way. The result was appalling. Great gaps were torn in the struggling mob, and the whole horde staggered and tore apart. Unsupported by the mass behind, the front ranks caved in, and over their mangled bodies we rushed into the narrow streets of Akka.

Opposition was not at an end. The stocky blue men still fought back. Up and down the streets sounded the clash of steel, crack of shots, and yells of pain and fury. But our greatest peril was from above.

The winged men were swarming out of their citadel like hornets out of a nest. Several hundred of them dropped swiftly down into Akka, swords in their hands, while others lined the rim of the cliff and poured down showers of arrows. Now the warriors hidden in the shrub-masked ditches opened fire, and as that volley thundered, a rain of mangled forms fell on the flat roofs of Akka. The survivors wheeled and raced back to cover as swiftly as their wings could carry them.

But they were more deadly in defense than in attack. From every casement, tower and battlement above they rained their arrows; a hail of death showered Akka, striking down foe and serf alike. Guras and Akkas took refuge in the stone-roofed huts, where the battling continued in the low-ceilinged chambers until the gutters of Akka ran red. Four thousand Guras battled four times their number of Akkas, but the size, ferocity and superior weapons of the apemen balanced the advantage of numbers.

Across the river Khossuth's carbineers kept up an incessant fire at the towers of Yugga, but with scant avail. The Yagas kept well covered, and their arrows, arching down from the sky, had a greater range and accuracy than the carbines of the Guras. But for their position among the ditches, Khossuth's men would have been wiped out in short order, and as it was, they suffered terribly. They could not join us in Akka; it would have been madness to try to cross the bridge in the teeth of that fire.

Meanwhile, I ran straight for the temple of Yasmeena, cutting down those who stood in my way. I had discarded the clumsy Akka sword for a fine blade dropped by a slain Gura, and with this in my hand I cut my way through a swarm of blue spearmen who made a determined stand before the temple. With me were Ghor, Thab the Swift, Than Swordswinger and a hundred other picked warriors.

As the last of our foes were trampled under foot, I sprang up the black stone steps to the massive door, where the bizarre figure of the Akka priest barred my way with shield and spear. I parried his spear and feinted a thrust at his thigh. He lowered the great gold-scrolled shield, and before he could lift it again I slashed off his head, which rolled grinning down the steps. I caught up the shield as I rushed into the temple.

I rushed across the temple and tore aside the golden screen. My men crowded in behind me, panting, blood-stained, their fierce faces lighted by the weird flame from the altar jewel. Fumbling in my haste, I found and worked the secret catch. The door began to give, reluctantly. It was this reluctance which fired my brain with sudden suspicion, as I remembered how easily it had opened before. Even with the thought I yelled, "Back!" and hurled myself backward as the door gaped suddenly.

Instantly my ears were deafened by an awful roar, my eyes blinded by a terrible flash. Something like a spurt of hell's fire passed so close by me it seared my hair in passing. Only my recoil, which carried me behind the opening door, saved me from the torrent of liquid fire which flooded the temple from the secret shaft.

There was a blind chaotic instant of frenzy, shot through with awful screams. Then through the din I heard Ghor loudly bellowing my name, and saw him stumbling blindly through the whirling smoke, his beard and bristling hair burned crisp. As the lurid murk cleared somewhat, I saw the remnants of my band-Ghor, Thab and a few others who by quickness or luck had escaped. Than Swordswinger had been directly behind me, and was knocked out of harm's way when I leaped back. But on the blackened floor of the temple lay three-score shriveled forms, burned and charred out of all human recognition. They had been directly in the path of that devouring sheet of flame as it rushed to dissipate itself in the outer air.

The shaft seemed empty now. Fool to think that Yasmeena would leave it unguarded, when she must have suspected that I escaped by that route. On the edges of the door and the jamb I found bits of stuff like wax. Some mysterious element had been sealed into the shaft which the opening of the door ignited, sending it toward the outer air in a rush of flame.

I knew the upper trap would be made fast. I shouted for Thab to find and light a torch, and for Ghor to procure a heavy beam for a ram. Then, telling Than to gather all the men he could find in the streets and follow, I raced up the stair in the blackness. As I thought, I found the upper trap fastened-bolted above, I suspected; and listening closely, I caught a confused mumbling above my head, and knew the chamber must be filled with Yagas.

An erratic flame bobbing below me drew my attention, and quickly Thab reached my side with a torch. He was followed by Ghor and a score of others, grunting under the weight of a heavy loglike beam, torn from some Akka hut. He reported that fighting was still going on in the streets and buildings, but that most of the Akka males had been put to the sword, and others, with their women and children, had leaped into the river and swum for the south shore. He said some five hundred swordsmen were thronging the temple.

"Then burst this trap above our heads," I exclaimed, "and follow me through. We must win our way into the heart of the hold, before the arrows of the Yagas on the tower overwhelm Khossuth."

It was difficult in that narrow shaft, where only one man could stand on each step, but gripping the heavy beam like a ram, we swung it and dashed it against the trap. The thunder of the blows filled the shaft deafeningly, the jarring impact stung our hands and quivered the wood, but the trap held. Again-and again-panting, grunting, thews cracking, we swung the beam-and with a final terrific drive of hard-braced knotty legs and iron shoulders, the trap gave with a splintering crash, and light flooded the shaft.

With a wordless yell I heaved up through the splinters of the trap, the gold shield held above my head. A score of swords descended on it, staggering me; but desperately keeping my feet, I heaved up through a veritable rain of shattering blades, and burst into the chamber of Yasmeena. With a yell the Yagas swarmed on me, and I cast the bent and shattered shield in their faces, and swung my sword in the wheel that flashed through breasts and throats like a mowing blade through corn. I should have died there, but from the opening behind me crashed a dozen carbines, and the winged men went down in heaps.

Then up into the chamber came Ghor the Bear, bellowing and terrible, and after him the killers of Khor and of Koth, thirsting for blood.

That chamber was full of Yagas, and so were the adjoining rooms and corridors. But in a compact circle, back to back, we held the shaft entrance, while scores of warriors swarmed up the stair to join us, widening and pushing out the rim of the circle. In that comparatively small chamber the din was deafening and terrifying-the clang of swords, the yelling, the butcher's sound of flesh and bones parting beneath the chopping edge.

We quickly cleared the chamber, and held the doors against attack. As more and more men came up from below, we advanced into the adjoining rooms, and after perhaps a half-hour of desperate fighting, we held a circle of chambers and corridors, like a wheel of which the chamber of the shaft was the axle, and more and more Yagas were leaving the turrets to take part in the hand-to-hand fighting. There were some three thousand of us in the upper chambers now, and no more came up the shaft. I sent Thab to tell Khossuth to bring his men across the river.

I believed that most of the Yagas had left the turrets. They were massed thick in the chambers and corridors ahead of us, and were fighting like demons. I have mentioned that their courage was not of the type of the Guras', but any race will fight when a foe has invaded its last stronghold, and these winged devils were no weaklings.

For a time the battle was at a gasping deadlock. We could advance no farther in any direction, nor could they thrust us back. The doorways through which we slashed and thrust were heaped high with bodies, both hairy and black. Our ammunition was exhausted, and the Yagas could use their bows to no advantage. It was hand to hand and sword to sword, men stumbling among the dead to come to hand grips.

Then, just when it seemed that flesh and blood could stand no more, a thunderous roar rose to the vaulted ceilings, and up through the shaft and out through the chambers poured streams of fresh, eager warriors to take our places. Old Khossuth and his men, maddened to frenzy by the arrows that had been showering upon them as they lay partly hidden in the ditches, foamed like rabid dogs to come to hand grips and glut their fury. Thab was not with them, and Khossuth said he had been struck down by an arrow in his leg, as he was following his king across the bridge in that dash from the ditches to the temple. There had been few losses in that reckless rush, however; as I had suspected, most of the Yagas had entered the chambers, leaving only a few archers on the towers.

Now began the most bloody and desperate melee I have ever witnessed. Under the impact of the fresh forces, the weary Yagas gave way, and the battle streamed out through the halls and rooms. The chiefs tried in vain to keep the maddened Guras together. Struggling groups split off the main body, men ran singly down twisting corridors. Throughout all the citadel thundered the rush of trampling feet, shouts, and din of steel.

Few shots were fired, few arrows winged. It was hand to hand with a vengeance. In the roofed chambers and halls, the Yagas could not spread their wings and dart down on their foes from above. They were forced to stand on their feet, meeting their ancient enemies on even terms. It was out on the rooftops and the open courts that our losses were greatest, for in the open the winged men could resort to their accustomed tactics.

But we avoided such places as much as possible, and man to man, the Guras were invincible. Oh, they died by scores, but under their lashing swords the Yagas died by hundreds. A thousand ages of cruelty and oppression were being repaid, and red was the payment. The sword was blind; Yaga women as well as men fell beneath it. But knowing the fiendishness of those sleek black females, I could not pity them.

I was looking for Altha.

Slaves there were, thousands of them, dazed by the battle, cowering in terror, too bewildered to realize its portent, or to recognize their rescuers. Yet several times I saw a woman cry out in sudden joy and run forward to throw her arms about the bull-neck of some hairy, panting swordsman, as she recognized a brother, husband, or father. In the midst of agony and travail there was joy and reuniting, and it warmed my heart to see it. Only the little yellow slaves and the red woman crouched in terror, as fearful of these roaring hairy giants as of their winged masters.

Hacking and slashing my way through the knots of struggling warriors, I sought for the chamber where were imprisoned the Virgins of the Moon. At last I caught the shoulder of a Gura girl, cowering on the floor to avoid chance blows of the men battling above her, and shouted a question in her ear. She understood and pointed, unable to make herself heard above the din. Catching her up under one arm, I slashed a path for us, and in a chamber beyond I set her down, and she ran swiftly down a corridor, crying for me to follow. I raced after her, down that corridor, up a winding stair, across a roof-garden where Guras and Yagas fought, and finally she halted in an open court. It was the highest point of the city, besides the minarets. In the midst rose the dome of the Moon, and at the foot of the dome she showed me a chamber. The door was locked, but I shattered it with blows of my sword, and glared in. In the semidarkness I saw the gleam of white limbs huddled close together against the opposite wall. As my eyes became accustomed to the dimness I saw that some hundred and fifty girls were cowering in terror against the wall. And as I called Altha's name, I heard a voice cry, "Esau! Oh, Esau!" and a slim white figure hurled itself across the chamber to throw white arms about my neck and rain passionate kisses on my bronzed features. For an instant I crushed her close, returning her kisses with hungry lips; then the roar of battle outside roused me. Turning I saw a swarm of Yagas, pressed close by five hundred swords, being forced out of a great doorway near by. Abandoning the fray suddenly they took to flight, their assailants flowing out into the court with yells of triumph.

And then before me I heard a light mocking laugh, and saw the lithe figure of Yasmeena, Queen of Yagg.

"So you have returned, Ironhand?" Her voice was like poisoned honey. "You have returned with your slayers to break the reign of the gods? Yet you have not conquered, oh fool."

Without a word I drove at her, silently and murderously, but she sprang lightly into the air, avoiding my thrust. Her laughter rose to an insane scream.

"Fool!" she shrieked. "You have not conquered! Did I not say I would perish in the ruins of my kingdom? Dogs, you are all dead men!"

Whirling in midair she rushed with appalling speed straight for the dome. The Yagas seemed to sense her intention, for they cried out in horror and protest, but she did not pause. Lighting on the smooth slope of the dome, keeping her perch by the use of her wings, she turned, shook a hand at us in mockery, and then, gripping some bolt or handle set in the dome, braced both her feet against the ivory slope and pulled with all her strength.

A section of the dome gave way, catapulting her into the air. The next instant a huge misshapen bulk came rushing from the opening. And as it rushed, the impact of its body against the edges of the door was like the crash of a thunderbolt. The dome split in a hundred places from base to pinnacle, and fell in with a thunderous roar. Through a cloud of dust and debris and falling stone the huge figure burst into the open. A yell went up from the watchers.

The thing that had emerged from the dome was bigger than an elephant, and in shape something like a gigantic slug, except that it had a fringe of tentacles all about its body. And from these writhing tentacles crackled sparks and flashes of blue flame. It spread its writhing arms, and at their touch stone walls crashed to ruin and masonry burst apart. It was brainless, sightless-elemental force incorporated in the lowest form of animation-power gone mad and run amuck in a senseless fury of destruction.

There was neither plan nor direction to its plunges. It rushed erratically, literally plowing through solid walls which buckled and gave way, falling on it in showers which did not seem to injure it. On all sides men fled aghast.

"Get back through the shaft, all who can!" I yelled. "Take the girls-get them out first!" I was dragging the dazed creatures from the prison chamber and thrusting them into the arms of the nearest warriors, who carried them away. On all sides of us the towers and minarets were crumbling and roaring down in ruin.

"Make ropes of the tapestries," I yelled. "Slide down the cliff! In God's name, hasten! This fiend will destroy the whole city before it is done!"

"I've found a bunch of rope ladders," shouted a warrior. "They'll reach to the water's edge, but-"

"Then fasten them and send the women down them," I shrieked. "Better take the chance of the river, then-here, Ghor, take Altha!"

I threw her into the arms of the bloodstained giant, and rushed toward the mountain of destruction which was crashing through the walls of Yugga.

Of that cataclysmic frenzy I have only a confused memory, an impression of crashing walls, howling humans, and that engine of doom roaring through all, with a ghastly aurora playing about it, as the electric power in its awful body blasted its way through solid stone.

How many Yagas, warriors and women slaves died in the falling castles is not to be known. Some hundreds had escaped down the shaft when falling roofs and walls blocked that way, crushing scores who were trying to reach it. Our warriors worked frenziedly, and the silken ladders were strung down the cliffs, some over the town of Akka, some in haste, over the river, and down these the warriors carried the slave-girls-Guras, red and yellow girls alike.

After I had seen Ghor carry Altha away I wheeled and ran straight toward that electric horror. It was not intelligent, and what I expected to accomplish I do not know. But through the reeling walls and among the rocking towers that spilled down showers of stone blocks I raced, until I stood before the rearing horror. Blind and brainless though it was, yet it possessed some form of sensibility, because instantly, as I hurled a heavy stone at it, its movements ceased to be erratic. It charged straight for me, casting splintered masonry right and left, as foam is thrown by the rush of an ox through a stream.

I ran fleetly from it, leading it away from the screaming masses of humanity that struggled and fled along the rim of the cliff, and suddenly found myself on a battlement on the edge of the cliff, with a sheer drop of five hundred feet beneath me to the river Yogh. Behind me came the monster. As I turned desperately, it reared up and plunged at me. In the middle of its gigantic slug-like body I saw a dark spot as big as my hand pulsing. I knew that this must be the center of the being's life, and I sprang at it like a wounded tiger, plunging my sword into that dark spot.

Whether I reached it or not, I did not know. Even as I leaped, the whole universe exploded in one burst of blinding white flame and thunder, followed instantly by the blackness of oblivion.

They say that at the instant my sword sank into the body of the fire-monster, both it and I were enveloped in a blinding blue flame. There was a deafening report, like a thunderclap, that tore the creature asunder, and hurled its mangled form, with my body, far out over the cliff, to fall five hundred feet into the deep blue waters of Yogh.

It was Thab who saved me from drowning, leaping into the river despite his

110

crippled condition, to dive until he found and dragged my senseless body from the water.

You will say, perhaps, that it is impossible for a man to fall five hundred feet into water and live. My only reply is that I did it, and I live; though I doubt if there is any man on Earth who could do it.

For a long time I was senseless and for longer I lay in delirium; for longer again, I lay completely paralyzed, my disrupted and numbed nerves slowly coming back into life again.

I came to myself on a couch in Koth. I knew nothing of the long trek back through the forests and across the plains from the doomed city of Yugga. Of the nine thousand men who marched to Yagg, only five thousand returned, wounded, weary, bloodstained, but triumphant. With them came fifty thousand women, the freed slaves of the vanquished Yagas. Those who were neither Kothan nor Khoran were escorted to their own cities-a thing unique in the history of Almuric. The little yellow and red women were given the freedom of either city, and allowed to dwell there in full freedom.

As for me, I have Altha-and she has me. The glamor of her, akin to glory, dazzled me with its brilliance, when first I saw her bending over my couch after my return from Yagg. Her features seemed to glimmer and float above me; then they coalesced into a vision of transcendent loveliness, yet strangely familiar to me. Our love will last forever, for it has been annealed in the white-hot fires of a mutual experience-of a savage ordeal and a great suffering.

Now, for the first time, there is peace between the cities of Khor and Koth, which have sworn eternal friendship to each other; and the only warfare is the unremitting struggle waged against the ferocious wild beasts and weird forms of animal life that abound in much of the planet. And we two-I an Earthman born, and Altha, a daughter of Almuric who possesses the gentler instincts of an Earthwoman-we hope to instill some of the culture of my native planet into this erstwhile savage people before we die and become as the dust of my adopted planet, Almuric.

www.ingramcontent.com/pod-product-compliance
Lightning Source LLC
Chambersburg PA
CBHW031729210326
41520CB00042B/1457